MATH

A Subtle Language of the Universe

(Raghavendra N Bhat)

First printing, 2019

ASIN: B07NYWBQL7 (Kindle eBook)

ISBN: 9781797672854 (Paper back)

First published by RNBhat Publications
37 Ananth Nagar, Manipal, Karnataka, India, 576104

Dedicated to my Father,

for believing in me and guiding me from age 0,

and to You, the Reader,

(Hoping that this book will be your first step towards viewing math differently.)

An equation means nothing to me unless it expresses a thought of God.

Srinivasan Ramanujan

Table of Contents

1. WHY IS MATHEMATICS TOUGH?.................................7
2. TEACHING MATH DIFFERENTLY.......................17
3. ANCIENT INDIAN MATH 26
   Geometry 36
   Mathematics in Coding.............................38
   Binary Numbers......................................40
   Magic Squares 42
4. SIMPLER TECHNIQUES FOR MENTAL MATH......47
   Addition..51
   Multiplication 55
   Calculating the day of the week for a given date 65
   Birthday Magic Square 68
   A note on subtraction: 77
5. WHERE DO WE SEE MATHEMATICS? 81
6. MATHEMATICS IN STORIES 92
   1. The Dilemma of the Traveler 93
   2. Scaling New Heights 96
   3. Soot for those in suits 98
   4. Efficient Nephew 100
   5. Lucky Student.....................................103
   6. Crossing the Fatal Bridge 105
   7. Pals on the plane.................................108
   8. Black and White on the Island................110
   9. Difficult Aliens...................................112
   Answers 115
EPILOGUE 120

1. WHY IS MATHEMATICS TOUGH?

Many things are easy, many things are tough;
Math is not tough, it is just a lot of stuff.

-Me

Before we get into the intricacies of this wonderful subject, let us first stop at something very basic. Let us ask a few questions that are seemingly obvious but have been asked for many a year by many a person.

1. Why is mathematics so hard? (for some)
2. Why is it boring? (for some)
3. Is it really boring? (For some)
4. Is it really necessary? (for everyone)

As a child, I was brought up by parents who planted a wonderful thought in my head: "Mathematics is the only subject where you needn't memorize anything." I was instantly fascinated!

To be precise, mathematics is the one subject that needs the 'least' amount of memorization as compared to its friends: biology, chemistry and social studies.

So why is it that something which requires the least amount of memorization is the hardest for so many wonderful people all around the globe ?
How can something that does not need memorization be boring ?

The mind, as it turns out, is designed in such a way that logic is something that is present in its innermost chamber. Does this metaphorically point to the fact that things which need logical thinking have to be thought long and deep? Is it hard *work*? Do people feel tired when they have to use logic? I believe it to be partly true.
Memorizing things and carrying out the timeless art of 'rote learning' somehow seems to be easier for so many people than actually digging deep into the canals of their brains and applying logic and analytic thoughts to a subject or situation. Humans always prefer the easier and shorter route, which has no thinking involved, if it yields the best and most productive results.

But what if it is a myth? What if 'applying logic' is not that hard? Maybe it is hard at the beginning. And once you train your mind to do it, the paths to access logic might become a lot easier. And comfortable. My experience with logic and math tells me that with time, the process of thinking and applying logic just keeps getting faster. It is like learning to tie a shoe lace. It is hard at first; but once your hands and motor nerves get used to it, the action just gets quicker and quicker till you can tie the lace without even noticing you are doing so. Applying logic to situations and training your mind for analytical thinking happens in a somewhat similar fashion. The more you do it, the more familiar the entire process becomes; and eventually you even end up doing it sub-consciously.

So what has this got to do with mathematics? Math, as it turns out, is a subject totally synonymous with logic. Without logic there is no math. The vice versa need not always be true however.

Almost every aspect of math has got to do something with logic and, more often than not, needs to be handled with a lot of analytical and multi-dimensional thought processes. A person would find it virtually impossible to master mathematics by just mugging up stuff. The positive part is that almost every thing in math can be explained in a very convincing logical way. So, those who always need justifications and explanations (like yours truly), will always be satisfied to work in most areas of math. The negative part though is that there is almost nothing in math that does not need the use of brain and logic.

What makes it worse is that if one *tries* to tackle mathematics with the rote learning approach, one will just end up digging oneself into a hole deeper than the one that initially existed. This is becase math becomes even harder and more irritating if you try to learn it without logic. It is like an animal that needs to be tamed. If you fight against it, its potential to harm you keeps increasing. However, if you try to understand its behaviour, it can, more often than not, end up being a wonderful friend.

So the main issue with math is the approach. Most people tend to have pre-conceived notions about math and tend to approach it wrongly. This can happen due to three reasons :
1. Someone they trust would have told them that math is hard and thus they already have a block in their mind about the subject.

This happens because *Homo Sapiens* have the tendency to blindly believe what others say without verifying it themselves.

2. They might have approached the subject in a wrong way to start with and never given it a second try with a different approach later on.

This happens because *Homo Sapiens* very rarely believe in giving second chances (be it to math or to relationships or to movies or to books).

3. They might have been taught the subject in a way that was not appealing (more on this later on in the book).

This happens because *Homo Sapiens* always believe in the 'first impression' theory. So if math is presented to them in an incorrect way to start with, the first impression will end up being very poor; hence causing them to instantly and permanently hate numbers.

Is math *actually* difficult? Well, everything can be looked at differently, based on one's perspective. Yes, it can be difficult; but is it as difficult as a normal person expects it to be? May be not.

There is a notion that there are two types of 'difficult'. For example, when someone says that running a 100 m race in less than 20 seconds is 'difficult', they are refering to one type of 'difficult'; and when someone says that trigonometry is 'difficult', they are refering to another type of 'difficult'. I firmly believe that the second type of 'difficult' is a mere myth and ceases to exist in a perfect world. Math is difficult. Yes. But it is difficult in the same way that running 100 m in less than 20 sec is difficult; or in the same way that getting up at 5:30 am is difficult. Math is something that each and every person in the world is totally capable of doing with ease and expertise if approached in the right way with a clear, open mind. It just works on three basic principles.

1. There is logic to everything (duh?)
2. Everything is connected
3. Everything is virtual but can be seen in reality (paradox?)

The second principle is also one of the reasons why many people find math difficult. Since almost everything is connected, if you mess up the basics it is very hard to catch up later on because learning every new concept is like adding a new storey to the building. Thus, without the proper foundation, it becomes almost impossible to add the storeys. The stronger the foundation, the taller the building can become. Something as basic as BODMAS (Brackets of, Division, Multiplication, Addition, Subtraction), which is taught to students no older than 12-13 years, can be highly important in concepts taught many years later.

Most parents do not realize that what they tell their child when it is young can have a great influence on their child's overall perspective of the world. It can shape the child's thinking, priorities, principles and beliefs in irreversible ways. Thus, when most children are told at a young age that math is hard (that too by their own parents), they will grow up with that seed of thought sown in their tender minds, that will, more often than not, stay firm even in their grown-up minds. Therefore, it is very important that even if a parent might have personally experienced troubles and complications with numbers, they ensure that they do not share those experiences with their children (especially when they are under 12).

When I was 8 years old, in the summer vacations of my 2nd grade in school, my parents made a small booklet full of problems ranging from 8-digit by 8-digit multiplications to 20-digit additions. When I look back, I feel shocked that the young kid in me did not find it appalling or cumbersome. This was mainly because of the sheer straight-forward and casual approach adopted by my parents.

They made it look totally normal. They made me believe that it was nothing more than just a normal 2nd grade holiday homework; and even if I did the problems, they hardly showed any amazement or surprise; rather they kept giving me more, longer and harder problems . The end result, though, was that an 8 year old boy believed that math was easy, an 8 year old boy had the complete confidence in himself that he could tackle the subject, and most importantly, it gave him a sense of joy everytime he got the answer right.

I was never frightened of getting the wrong answer. I was rather looking forward to the joy of getting a 16-digit answer and it being 100% right.

This might sound very cliche, but numbers do tend to become friends if you understand them and love them (yes, I know you have heard this many a time and always felt 'what rubbish'). But it really is true. If you spend more time with them, you begin to connect to them intuitively. You start seeing numbers as lyrics of a beautiful song. The sheer magnificence of them to always give you the same answer whether you do it this way or that; the consistency by which the answer remains the same in 200 BC or 2019 AD; the mystery by which some of them are not explainable (the primes, for example); the endless nature of them; are just a few of the magical traits of numbers.

But, almost everything in math is virtual. It all happens in the mind or on a piece of paper. This is why many people find it 'beyond the realm of thought'. For example, people find it easier to connect to the study of the human body or the study of ancient civilizations because they feel it is more *real*. They can see it with their own eyes, they can hear it with their own ears and, more often than not, there is actual physical and practical proof for most of these topics' existence. They feel that math is a lot of 'hocus pocus' that is unnecessary and does not exist. I just want to ask a couple of simple questions: ***What if you don't know where to look? What if it exists for real and you haven't searched enough for it ?***

As you go further into this humble work of mine, you will begin to realise that math *does* exist in real. It is infact the language in which the book called "The Universe" is written. It gives an explanation for almost every thing that existed, exists or will exist. But the problem though, is that it is a hidden language. It is like Hieroglyphics. It has to be carefully studied and decoded. It can be a tedious process. But once done, the results are unbelievable!

Most people feel that the theoretical part of mathematics - involving proofs and facts and figures - is boring. Yes, it can sometimes be. There was a time when I felt that the most redundant part of mathematics is the process of proof-writing. But I realized later on that math is the only subject where something can actually be proved for real.

In math, if something is proved once, it stays. Forever. Something proved by Euclid in the 3rd century BC (such as the proof of the existence of infinite prime numbers), absolutely holds true even today. In other subjects like the sciences, proofs are mostly temporary. There is always scope for improvement, there is always a contradiction and there is always a topic for debate. Something which was proved to be true a few years ago, can be rewritten and disproved by a new scientist. The world would have to rewrite many text books, and everybody would have to adjust to this new thought process. There is no ultimate truth; discoveries keep happening; 'facts' keep changing. It was initially believed that the liver could be pumping blood; later on, it was *proved* that the heart pumps blood; and one cannot even predict the gravity of future theories on the same. I do not mean to say that the sciences or the arts are unstable. They are all vital to our existence just as mathematics is and are all perfect in their own right. They just do not tend to have something that '**will always be true no matter what**' or something that is '**100% real and proved**'.

Mathematics however begs to differ. Every discovery ever made, every paper ever written, every theorem ever proved just adds storeys to the universal 'Math Building' and no matter who says what, this building can never be demolished, rebuilt or reshaped. It can just become taller, stronger and sharper, and more well-defined and more beautiful. Even the smallest of math discoveries made will stand the test of time, and even the most basic proof ever written or the 'silliest' of axioms and postulates will always be true. Forever.

When mobile phones first came into the market there were many people who believed that they are unnecessary and worthless. When Galileo first shared his theories about the universe, people laughed at him. When Mahatma Gandhi first shared his beliefs and methods people felt he was a fool. However, with time, people ultimately succumbed to the actual worth and merit of the above mentioned. This tells us something about human psychology. Every person has an ego, a belief, a pre-conceived notion and one's own version of 'the ultimate truth'. But every person also has a potential to *suppress* one's ego, *change* one's belief, *rethink* on one's pre-conceived notion and also to *reshape* one's understanding of what is 'real and true'. You, the reader, who started reading this book assuming that math is hard and tedious, I sincerely hope that you can give it a second chance and approach it differently because, frankly, if you do not, you are missing out on loads of fun!

2. TEACHING MATH DIFFERENTLY

If the student has not understood,
The teacher has not taught.

-Anonymous

The most famous talk site in the world is TED.com; and one of their most viewed talks is by Dr. Ken Robinson: "Do schools kill creativity?" It has been decades now that this wonderful education system has existed on Earth. It is nothing short of a global phenomenon and it has resulted in employment, development in science and technology and also paved the way for the growth of artificial intelligence.

Why then is the most viewed talk of the most famous talk program in the world about the negative effects of the education system on the creativity of children? Is there actually something huge that we are missing out on? I firmly believe that nothing in this world can be perfect. Everything will and should have always have a positive side and a negative side to it. The debate then, should only be about which of the two sides, positive or negative, dominates.

The prevelant education system definitely has more positives. It is without doubt something truly magnificent.

Though I have my own share of complaints against schools, it would be foolish to say that my school never helped me to become what I am now or what I will become ten or even twenty years from now. Where does the problem lie then? Can we hold the education system responsible for the aversion towards math worldwide? Can we hold the style of teaching responsible for the so called 'fear' of numbers? Well, not solely; but partly for sure.

The problem with mathematics is that it is very different from almost all other subjects. Therefore, it goes without saying that the style and approach used towards teaching mathematics has to be different from anything else ever taught. The biggest and most vital branch in this tree of math teaching should be **C**arefully **R**especting, **E**fficiently **A**pproaching and **T**errifically **I**gniting **V**ital **I**ntellectual **T**houghts in **Y**outh. (CREATIVITY)

Dr. Robinson says, "Creativity now, is as important in education as literacy, and it should be treated with the same status." In the case of mathematics nothing could be more appropriate. The entire field of mathematics is a result of the thoughts of millions of creative minds over the course of several centuries. Creativity is the *art* of having one's own ideas and thoughts. Math is all about remarkable ideas that have meaning.

So how do you make sure that a person thinks creatively? Do all people think creatively? Every single person is born with a mind that has potential for creative thought processes. The challenge is to allow the mind to grow in an environment that will appreciate and nurture that wonderful ability.

Human psychology tells us that most people become insecure and unsure of themselves when their actions are judged as wrong. People shy away from doing things if they are worried about the outcome. Most children are brought up in an atmosphere where they are frightened of being wrong. Imagine a question which has neither right nor wrong answers. Almost every child will feel the freedom to give an answer to such a question. Nobody will ever have the fear of making a mistake. Now extend this imagination to a classroom which accepts all responses, never branding a student as right or wrong. Every student will actively participate in such a learning environment, without fear. What actually happens then, is that since the children will not be frightened of being wrong, they will be motivated to try. If someone gives you a parachute and assures you that you will not get hurt , you will fearlessly jump from any height. If somebody gives you a pair of floats, you will not fear drowning and will not shy away from even the deepest of seas. That safety net or that sphere of encouragement can only be given by the teachers and parents of the child. A mistake is nothing but an assurance that someone is actually trying. For somebody to get a 'wrong' answer, they *have to have an answer.* And the very fact that they have an answer means that they were brave enough to try.

Mathematics is a subject where there is no fixed rule to solve a given problem. It is like life. When you face a tough situation in life, there is no specific rule to apply to tackle the situation in a particular way. You may approach your difficulties in whichever way you want. You may meditate, or approach a friend, or read a book or even watch a movie. The only thing that really matters is the actual improvement of the situation and the resolution of the problem. Math works in a similar way. The answer is all that matters. The approach need not and should not matter.

I believe that everyone may have his/her own comfortable way of solving a problem. One should be given that freedom. With this freedom, one may sometimes come up with new methods to solve already solved problems; and these new methods might end up being easier than the existing methods. Sometimes, the new methods might help solve hereto unsolved problems.

The mind works in a different way when it is told to 'do whatever you want, just get the right answer' as compared to 'use only this method to get the right answer'.

In the first approach, the mind is given that space, that trust and that freedom to explore and think. The mind then uses its creativity and its multi-dimensional thought process to solve the problem.

In the second approach however, the mind becomes a mechanical machine; more like a computer or a humanoid robot. The mind then stops using its own creativity and thought processes and gets adjusted to being given word-to-word instructions and becomes dependent on external means to be able to think. This will eventually become a habit and when left all by itself (say, in an exam or a real life situation), the mind will feel weird and out of place. The human mind adjusts to instructions the same way that a human body adjusts to medication. After a while, such instructions, or medication, become a habit and more importantly, a necessity. The mind gets used to being 'spoon fed'. A creative mind however, will have the confidence in itself to tackle any given situation because it would already be trained to think independently.

The other thing about mathematics is that it becomes fun when there is discussion and interaction between different schools of thought. The mind works best when it is lost in deep conversation. The brain prefers company to loneliness. If students are allowed to discuss and debate, the range of their creative thinking just increases. Sometimes in classrooms, students are made to work individually. Some students might not be confident to work on their own. Maybe they would be more comfortable to solve problems as a group. Yes, ultimately everybody should be trained to think individually but it need not always be the first approach adopted.

There is something else that math is comprised of. It is the single most vital component that is often ignored and suppressed in most schools: **'The art of asking and answering questions'**. A normal human mind generates about 2500 thoughts every hour. This means that the mind is designed to continuously think thoughts and wonder about stuff. It is of course impossible to ask questions arising from each of those 2500 thoughts; but something that can greatly help the classroom culture is the freedom to ask questions. When a student asks a question, there is only one intention: to find the answer. There is no selfish motive; it is not an attempt to prove oneself to anybody; and the student will not gain anything else other than assistance in finding the solution to the query.

Over time, in such a classroom, students will become confident to ask questions. They will never be afraid of people; they will not have a problem to face a crowd; and they will have a lot of confidence in their own perspectives as well as train of thoughts. Since most classrooms have same aged students, an answer to one student's question might actually be an answer to thirty students' (size of the class) questions.

If suppressed however, the growing mind of the child could be negatively affected. It could feel less confident; or feel insecure; or worse still, have the constant fear of being wrong. This can have tremendous after-effects on creative thinking. Someone who is afraid to ask a question may also have the tendency to be afraid to try something new, or afraid to be creative or original.

In September 2017, on my trip across the United States, I made friends with some Indian parents who wanted me to tutor their children. Teaching is something that I have always enjoyed and practiced for over three years now. Something that I have noticed is that most children do not like to take tests. So I have made it a point to have very few tests and examinations for my students.

Children do not like tests due to several reasons:

1. They do not wish to be compared to their peers.

2. They feel it unnecessary to prove to the teacher that they do know or do not know the stuff.

3. They might not all have the same levels of attention span to sit at a place for hours to attempt question after question.

The first reason is the most valid complaint against tests. People do not like being compared to fellow beings. They wish to be respected for their individuality, appreciated for their abilities and helped overcome their short comings. Tests create a sense of superiority among a few and inferiority among many. The actual idea behind testing is pretty noble. Tests are a way to check if the student has actually understood what was taught.

This can be done through other, more productive ways as well. For instance, a set of questions can be given to every child and the teacher can tell the children to do those problems whenever they wish, however they want.

Later, on a decided date, the answers can be discussed by the class as a whole. This way, nobody would ever feel inferior or superior to any other person. It will just educate everybody about their mistakes and confirm their strengths.

The whole concept of awarding marks or points just creates a 'competition' feel to something that is not even a race or contest. Learning is something that should happen at an individual level and have positive influences on each and everyone involved.

Another alternative is to conduct the tests as usual, but keep the marks secret. But even then, telling a child that he/she is only five out of ten, can be demotivating.
Confidence is highly important for creativity. Without self-belief and faith in oneself, it is very difficult to be creative. Examinations and tests, more often than not, may create an identity crisis or inferiority complexes in children (especially those under 15).

I have great respect for the education system and I am forever indebted to my teachers for all the positive influences that they have had on me. I do not mean to undermine any individual or any system. I am just suggesting alternatives based on my understanding of math, feedback from my students and also from a few personal experiences. Since math is all about brain and logic, it is of utmost importance for the learners to have a stable mind-set. Therefore, a proper understanding of child psychology is as important as proper understanding of math concepts if one wants to become a math teacher.

Simple theorems and ideas need to be left open for discussion. When children are left to themselves, and given the freedom to explore, they will, more often than not, arrive at the answers by themselves. This will be a lot more efficient than actually spelling it out to them. One will never forget something if one has explored and discovered it on one's own.

Introduction of puzzles and math games in a math class is another way to instill the interest and passion for the subject (more on this in the last chapter). When people find something fun and entertaining, they like it. When they like it, they will never find it unnecessary or boring. The whole idea behind teaching math to children is to do it in as entertaining a way as possible; thus ensuring that they never quit, even if it might become difficult or challenging. If they are always interested, they will have the 'never give up' attitude.

And honestly, what more would a teacher want?

3. ANCIENT INDIAN MATH

East or West, Look within for the best.

-Me

The great thing about mathematics is that there are various ways of arriving at the same correct answer. There is only one correct answer but there are numerous correct approaches and paths to reach that answer. Over the years, many countries, cultures and civilizations have invented subjects that have been vital for what the world is today. The beauty of the discovery of math across the world, though, is that it has happened in different ways, at different times, in different places. Yet, ultimately, today, the world uses more or less the same math concepts and almost every country teaches similar math concepts to its students. Some have been inspired by nature, others by the need to survive. The pre-historic man counted on his fingers and for him, the number 'ten' was probably the upper limit of his counting range. The need to count might have risen to keep track of his weapons and animals. As time progressed, need and greed increased and also the necessity to have a well-defined number system.

The problem with many schools of thought is that there is no written record or proof of their discoveries. Thus, it makes it very difficult to study their minds or creations in detail. Indian history however, does not have this particular short-coming. There are many mathematical texts and scriptures written in 'decipherable' languages that make it easy for modern day historians to study them. Mathematics, as it turns out, was a staple diet for ancient Indian scholars.

The main intention of this chapter is to look at some very basic (and hopefully easy) math concepts and understand how they were developed in ancient India. As an Indian, it is a matter of great pride for me that some crucial mathematical discoveries and inventions happened in India. They have also been convincingly explained along with their applications, in ancient Indian mathematical texts. Whether the discoveries and inventions first happened in India is a matter of historical survey, but what matters is that they happened independently with respect to any other civilization and do not seem to have been borrowed from, or inspired by, any other school of thought. The originality, however is commendable.

Most civilizations had already used numbers for the purpose of counting, but converting numbers into symbols and actually writing them (on walls, palm leaves, etc.) was something that ancient Indians indulged in.

Before we get into the actual discussion on ancient Indian mathematics, there is a very important clarification to be made. There are two seperate terminologies : 'Vedic mathematics' and 'ancient Indian mathematics'. 'Vedic mathematics' is a book written by Swami Bharati Krishna Tirtha[1] in 1965. It contains a list of mental calculation techniques *claimed* to be based on the Vedas. However, it is not the same as 'ancient Indian mathematics'. When most people say 'Vedic Math' they refer to the book (sometimes unintentionally). Most people are also of the opinion that ancient Indian math is all about quick arithmetic (again, owing to the confusion between the book and the actual subject).

What we will be discussing is how math developed in ancient India and how Indians of long ago approached mathematical concepts which most of us are aware of today. We will not be focussing on mental math (not in this chapter at least).

Firstly, it is important to understand the mindset of the people living in ancient India. Today, a great mathematician is known for his expertise in mathematics; a great musician is known for his performances and achievements in the field of music; and a great author is known for his best sellers. However, in ancient India, more often than not, there were no 'single field' masters. There was hardly any human being who could be labelled as a 'Master of one'. Neither was there a 'Jack of all'. Most of these scholars were 'Masters of all'.

1

Jagadguru Swami Sri Bharati Krsna Tirthaji Maharaja, Vedic Mathematics, Motilal Banarsidass Publishers, Delhi, revised Ed (1992), 2001.

A great mathematician was more than likely to be well versed in literature; would know how to look at the sky and predict the next eclipse; would be a composer of many poems; and sometimes would also be a good presenter of songs written by him. The literary part is of utmost significance here because almost all theorems and proofs were written in the form of 'Shlokas' (short poems) in ancient Indian languages such as Sanskrit.

Every Shloka would be written in such a way that it would be grammatically correct, and would seldom miss rhyming patterns. All of this tells us that, in ancient India, for a person to be a mathematician, literary pandit, astronomer, composer and musician, an all in one, was not something very uncommon. Very rarely would you find a person excelling in only one art form; typically they were all *polymaths*. This is the prime reason that many subjects are interlinked in the ancient Indian scriptures. This also meant that each topic's applications across multiple fields was well understood.

Some of the finest mathematical discoveries happened more out of belief and necessity rather than motivated by a 'quest for glory'. Ancient Indian scholars were perfectionists of the highest order. The only way to ensure perfection of any sort was to apply mathematics. Thus, they had no other option but to resort to mathematics as their premier tool. Let us explore some concepts of math that have been explained in detail in some popular ancient Indian texts.

Concept of Zero

(The concept of Zero is vast. For the high-profile, spiritually and mathematically advanced minds reading this book, I wish to apologize to your high intellects for the basic concepts explained here. This book is mainly aimed at giving a broad and basic perspective about math concepts to any person willing to try to understand and love math)

Many people are of the opinion that the 'zero' was invented in India. This is said to be the most famous mathematical discovery of ancient India (need I even say the famous one liner: "India's contribution towards mathematics is 'nothing'"?). To say that the 'zero' was actually 'invented' in India is only a conjecture; but to say that it was 'discovered' in India would not be wrong at all.

A very common myth is that the great mathematician Aryabhatta discovered 'zero'. The truth, however, is that the idea of 'zero' and the concept of 'nothingness' existed in ancient India even before Aryabhatta.

This inconsistency happens when there is a difference of opinion about 'What is zero?'

'Zero', as it turns out, has so many different interpretations. Let us look at the simplest meanings and try to briefly study them.

1. **The concept of Shoonya**: The first thing that comes to our mind when we hear the word 'zero' is the concept of 'nothingness'. This concept was broadly explained by the great Indian Mathematician Pingala in the 3rd century BC. Many civilizations have simultaneously come across the concept of 'nothingness' and many of them might have stumbled upon it without intending to. They might have been in pursuit of discovering and working on something without knowing that 'that something' was actually 'nothing'. This 'nothingness' meaning of 'zero' is the first which is taught to children. It is something which is mostly used while counting. It is also probably the most common definition of 'zero' used in literature and language, where 'zero' is actually used as a word to represent 'nothing'. (For example, **'He has zero knowledge about astronomy'**).

In ancient Indian astronomy and spiritual thought, the beginning of the universe is believed to have happened from virtually nothing (science now talks about this as the Big Bang). So, 'Shoonya', or 'nothing', can be used to represent the beginning of the universe. As I stated earlier, a scholar in ancient India was well versed in many aspects of life; thus it was natural for them to connect concepts across fields.

It remains purely a conjecture that the pre-historic man was aware of the 'nothing' concept. Suppose he was using his fingers to count his animals everyday. What would he do if he had no animals to count? Would he realize that the concept of 'not having any animals' can also be represented by a symbol? One can only wonder. We now know that while enumerating all the permutations and combinations of doing something, doing *nothing* is also one of the many ways of doing *something*.

2. **Zero as a number:** When we actually write 'zero' in math on paper, the symbol is used as a *number*, or more specifically, as a *digit*. This concept of 'zero' came much later as compared to the Shoonya concept. When people say Aryabhatta discovered 'zero', they actually refer to this particular definition of 'zero' (sometimes purposely, sometimes inadvertently). Aryabhatta has used the digit 'zero' in his works; so it would not be wrong to assume that it was a concept fairly well known by then. So, we can broadly say that the concept of the digit and the number 'zero' happened around Aryabhatta's time. It is very difficult to say with certainty or find out who the first ever person who used this idea could be. It is similar to finding out who invented the English language. These are broad concepts that took shape over time and gained clarity over every passing generation. It would be incorrect to assume that one fine day a person woke up and decided to use this concept.

Since it is only conjectured that India was the first mathematical civilization to develop actual usage of 'zero' as both a number and digit, I will choose to say that **"Indian mathematics has definitely explained and used**

'zero' as a digit very long ago and if not the first, it is definitely one of the first ever cultures to do so."

Using 'zero' as a number is very useful for the birth and development of many areas of mathematics. The human being has ten fingers. But before the discovery of 'zero', there were only nine prominent symbols in the number system. With the addition of 'zero', it gave a more 'whole' feeling to dealing with numbers. Humans understood 'ten' better than 'nine'. It made math and counting internalizable.

Firstly, it led to the development of the **Place Value system**; the importance of which has been clearly noticed in ancient Indian texts. The 'Place value' idea is to write digits (symbols) in such a way that every digit has a specific *worth*, based on the position where it is placed. Thus, a digit's 'place' in a number decides its 'value'. and thus the name : Place value system. This was initially designed in 'base ten', which meant that numbers were counted in multiples of ten, and each digit's worth increased in multiples of ten, going from right to left.
Indian mathematicians thus used 'zero', the digit, to develop the place value system which facilitated mathematical tasks such as addition and multiplication with tremendous ease as compared to earlier methods (such as the Roman Numerical system). With the advent of 'zero', large numbers could now be written on paper. Numbers which could only be imagined before, could now be

multiplied with other large numbers to arrive at even larger numbers.

Thus, 'zero' became this magical digit that could be used to write large numbers as sums of powers of ten. The 'zero' became this amazing invisible tool that made calculations fairly easy and understandable for the common man.

Secondly, the 'number zero' gave rise to the concept of negative numbers. The 'zero' became this border that separated the positive numbers and negative numbers. It became the reference base to increase or decrease endlessly on either side. Thus, 'zero' made it possible to have infinitely 'more' as well as infinitely 'less' quantities at the same time. The great scholar Brahmagupta was one of the first Indian mathematicians to explain the concept of negative numbers. He explained this concept through the idea of 'owing' something to somebody, or the idea of 'debt'. This was something which was lesser than 'nothing'. This is how the idea of negative numbers took origin.

Negative numbers in today's world are applied almost everywhere. For example, a quantity like temperature can not be represented without negative numbers. Same is the case when discussing concepts like growth and decline. These are concepts that can have situations where 'zero' has to act as a neutral base line, rather than behaving as 'nothing'. Therefore, here, 'zero' refers to a pre-decided mid-point that separates the infinite positive numbers and the infinite negative numbers.

This may also be explained through the concept of giving answers to a question:

giving the right answer is 'positive';

not giving any answer is 'zero';

giving the wrong answer is 'negative'.

3. Zero to represent the circle of spirituality: As this is something that is more spirituality than mathematics, I would just like to say that 'zero' was used to explain the spiritually deep concept that beginning, ending, nothing and everything are all basically the same. Here the shape of 'zero' is the subject of discussion, i.e, the circle shape. This circle is a symbolic representation of life: a circle has no beginning or end, but is just a collection of infinite points. Thus, the circle is said to represent the ambiguous nature of existence that has neither beginning nor end. No one can just look at a circle and predict with certainty as to how it was drawn and which was the actual starting point. This concept of zero as both the beginning and the end also connects it to another mathematical concept: 'infinity' or endless. This is beautifully explained in a Shloka :

ॐ पूर्णमदः पूर्णमिदं पूर्णात्पुर्णमुदच्यते
पूर्णस्य पूर्णमादाय पूर्णमेवावशिष्यते ॥

**Om poornamadah poornamidam poornaat
poornamudachyate
Poornasya poornamaadaaya
poornamevaavashishyate ||**

This, in brief tells us that both zero and infinity are unique
because if you add or subtract them from *themselves* you
are ultimately left with *themselves* either way.
[0 + 0 and 0 − 0 both gives the answer as 0; $\infty + \infty$ and $\infty - \infty$ both gives the solution as ∞]

Zero and infinity are both refered to as 'Poornam' which
means 'complete'. A circle is complete and has infinite
edges and points in it. Almost all celestial objects (the Sun,
planets, stars, etc) appear spherical in shape and revolve
around other celestial bodies in orbits that are circles and
ellipses. Thus, the shape of the symbol used to represent
'zero' is spiritually and astronomically very significant.

Zero also has poetic metaphors such as 'Nothing by itself
but everything when put next to another'.

Geometry

Ancient Indian mathematicians were very passionate about Geometry. When we talk about applications of math in daily life, geometry is the most practically visible application-based concept among almost all math concepts. It was called Sulva Sutras. 'Sulva' means 'rope'. Basic geometry in India was developed just with the help of a rope. It is indeed a matter of awe that complex concepts such as trigonometry, mensuration and also study of various types of two-dimensional as well as three-dimensional shapes have all been explained in the ancient Indian texts. In ancient India, geometry took birth because of certain religious necessities. There was belief that a particular shape of the 'homa kundam' (fire altar) used in the religious formalities would appease a certain deity . Thus, they had various types of shapes for the 'homa kundas' depending on the deity being appeased. Some shapes were also inspired by animals, such as the falcon.

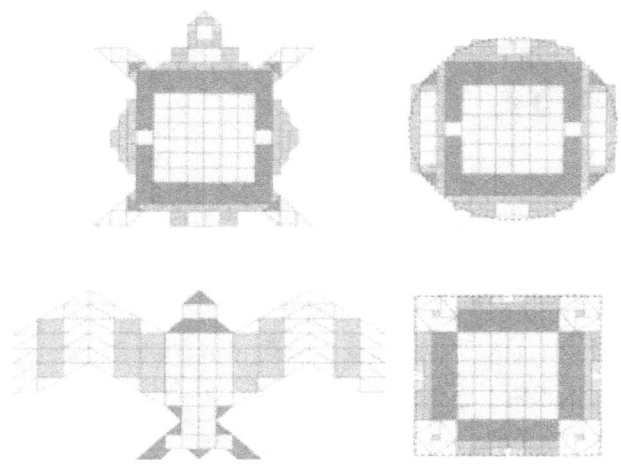

As noticeable from the picture, these required precision of the highest order. Moreover, the ancients believed that any mistake of even the smallest magnitude would unleash the wrath of the deities. Hence, they were very careful and their error percentages were literally negligible.

Brahmagupta, again, was a stalwart in this field. He has elaborated many mathematical concepts such as calculation of areas of triangles and quadrilaterals. His major area of interest was cyclic quadrilaterals (quadrilaterals which can be inscribed within a circle).

Mathematics in Coding

Arguably the most ingenious application of mathematics and numbers was in the field of cryptography and coding. Ancient scholars adopted various coding techniques to conceal numbers. One such technique was to convert them into words. Thus, only a person who knew of the coding technique would be able to decode the word. For a common man, it would just seem to be a normal word without any specific inner significance.

One such coding technique is the 'katapayadi' coding system. It assigned different numerical values to each syllable of the ancient Indian Alphabet. For example, a 4-digit number could be converted into a four letter word. However, since most mathematicians were also poets and literary geniuses, they would somehow make sure that even if they converted a number into a word, the word would be meaningful. They would creatively weave magical poems and meaningful texts that would actually have hidden numerical meanings beneath the apparent textual meanings. One such example is the 'pi song' from ancient India (it has been referred to in Swami Bharati Krishna's book as well). Pi, the irrational number (3.14 or 22/7 as used in day-to-day math), has fascinated mathematicians since times immemorial. The very fact that it is irrational, never ending and very mysterious, made many mathematicians spend years on de-mystifying Pi. This verse is one of the most intelligent ways to represent Pi.

गोपीभाग्यमधुव्रात-शृङ्गिशोदधसिन्धगि ।

खलजीवितिखाताव गलहालारसंधर ॥

**gopi bhagya madhuvrata srngiso dadhi sandhiga
khala jivita khatava gala hala rasandara**

This is actually a short poem written in the praise of Lord Krishna. However it is crafted in such a way that if decoded using the 'katapayadi' encryption technique (representing each syllable with a corresponding digit), it can be deciphered as:

31415926535897932384626433832792

All we have to do is place a decimal point after '3' and we get the value of Pi upto thirty decimal places! This can give birth to several questions. "How did the author calculate the value of Pi correctly up to so many decimal places?" "How did he find words such that they aligned in a perfectly logical and meaningful poem and also simultaneously encoded all the digits?"
One can only wonder at the genius of the author's mind!

Coding techniques were used by kings, astrologers and also spies to share confidential numbers amongst themselves.

Binary Numbers

In the present world, computers are probably the most important components in day-to-day life and also for the growth and maintenance of the world's terabytes of data. For the existence of computers and for the development of many other technological devices, binary number system (mathematical calculations using 0s and 1s by following the base 2 number system) is the most basic and important foundation. The entire structure of a computer is built with the help of binary numbers. The modern binary number system was studied in Europe in the 16th and 17th centuries.

It is astounding that the great Indian mathematician Pingala has given references to Binary numbers in his text, *'Chandaḥśāstra'* which was written in the 3rd century BC. It is important to note however, that what Pingala explained was not exactly the same as the binary number system that we understand and use today. The main ideology was, however, strikingly similar. Since the numerical system with 1 and 0 was not very prevelant at that time, the binary concept was explained through sounds and alphabets.

Short sounds (known as Laghu in Sanskrit literature) represented '0' and longer sounds (known as Guru) represented '1'. Various combinations of these sounds were used in poetry. Pingala gives detailed rules for listing all the possible permutations and combinations of short (0) and long (1) syllables to create words.

This concept was later developed by other mathematicians and poets (such as Acharya Virahanka and Acharya Hemachandra) to create and logically explain certain musical concepts. (More on this later in the 'application of math' chapter).

This great work of Pingala ('**Chandaḥśāstra**') also acts as one of the first proper concrete developments in establishment of Sanskrit grammar. What is even more astounding is that Pingala also uses the word 'Shoonya' in his works. However, the concept of 'zero' was very new at that time and it is hard to understand which meaning of 'zero' he was actually referring to.

Magic Squares

Magic squares, sudokus and other puzzles are those math-related games that are designed to be entertaining as well as intellectual. These wonderful boxes are very popular in mathematics because they can be appealing to a 'non-math' person as well. A magic square, in simple words, is a box created in such a way that if any row, column or diagonal is added across the box, the sum total would be the same. This sum total is called the 'magic sum'.

In Indian history, magic squares have been found in texts dating back to the 1st century AD. The strange thing however, is that magic squares were found first in texts that were more focussed on astrology than mathematics.
The 'three by three' magic square first appears in India in a book titled '*Gargasamhita*' by Garga. This book was written mainly on divination (predicting the future through super-natural means). This book also happens to have the first written records of magic squares in almost all recorded history of any civilization in the world. Garga referred to a 'three by three' magic square as a symbolic representation of the 'navagraha' (nine planets of the solar system : 'nava' means 9 and 'graha' means planet).
It is truly amazing that something that is purely arithmetic and hardly anything more than addition and arrangement of natural numbers was once used in astrology. This could once again be the result of the 'master of all traits' characteristic feature of ancient scholars.

Another reference to magic squares happens in a medical text, '*Siddhayog*' written in the 10th century AD. Surprisingly, this book was aimed at women undergoing labour.
References to magic squares are also found in other books on chemistry, and alchemists' texts.

The great ancient Indian scholar Varahamira wrote a book titled '*Brahatsaṃhitā*' (6th century AD).

Here we see one of the first written records of the 'four by four' magic square (which has sixteen boxes or cells). Here, a magic square was constructed for the purpose of creating perfumes (yes, you read that right) using sixteen different ingredients. Each of the sixteen cells represents a particular ingredient, while the number in the cell represents the proportion of the associated ingredient, such that the mixture of any four combination of ingredients along the columns, rows, diagonals, and so on, gave the total volume of the perfume-mixture to be 18.

Thus, the property of the magic square to give multiple answers to a single question was a useful tool for the development of many sciences.

Most magic squares are designed such that every row, column and diagonal in the box all add up to the same number. These ancient magic squares from the 'Brahatsaṃhitā' were designed in such a way that many other combinations of numbers inside the box gave the same answer in addition to rows, columns and diagonals.

Magic squares were found not only in ancient texts but also in the most unexpected of places, such as temple walls. Below is an image of one of the walls of an Indian temple (the Parshvanatha temple which was built in the 10th centry AD).

The numbers are written in a language from the past. When translated to the modern day number symbols, we get:

7	12	1	14
2	13	8	11
16	3	10	5
9	6	15	4

This turns out to be a 'consecutive number' magic square with numbers ranging from one to sixteen. It is designed in a unique way to give the same answer as many times as possible. There are more than thirty different ways of arriving at the magic sum of thirty four.

The question that arises at this point is 'Why did they need magic squares and why did they trace them on temple walls?' Temples are places of spirituality and since most ancient scholars were staunch believers of God and rituals, it is very likely that magic squares had astrological as well as spiritual meanings.

One more amazing discovery was when some ancient Indian coins were found having numbers divided into boxes on one of their sides.

Hence, there are written records of magic squares having applications in chemistry, medicine, commercial markets, astrology as well as economics. A few years ago I found a way to apply the properties of a magic square in developing music (more on this in the 'Application of math' chapter).

Thus, we conclude that Indian mathematicians had developed many modern day math concepts using original and different approaches, sometimes even before most other civilizations (as far as we can tell). The mindset and thought process of those ancient scholars was very different from what we expect to find in today's mathematicians. The biggest question still remains, 'How?' Many modern day researchers feel that there was divine intervention and most discoveries happened as a result of meditation and severe penance. We can only wonder. The remarkable thing though, is that there is proof and proper evidence for most of their discoveries.

It might not be possible for us to find out the motive, method or manner used by our ancestors but we can surely be inspired to live life in 'different' ways and explore various schools of thought through our own perspectives.

4. SIMPLER TECHNIQUES FOR MENTAL MATH

***Magic** is believing in yourself. If you can do that,*
You can make anything happen.

-Anonymous

Most people are afraid of mathematics because they find it
hard to do it in their head. They struggle to do simple math
in their head and ultimately rely on calculators and paper to
solve even the most basic of math problems. As time
progresses they get accustomed to calculators and feel
handicapped without them. One of the best ways to make
math easy and enjoyable is to do it mentally. The people
who know how to do fast mental math seldom hate it; nor
do they find the subject difficult. On the contrary, there are
very many people who do not know how to do mental math
and find almost everything about math daunting. These
people end up hating the whole subject altogether.
Thus, many people dislike math not because of the topics or
the subject; but because they feel it is beyond their thinking
sphere to master it. However, when they know that they can
do it mentally, most people find it a lot easier and
eventually it becomes like any normal thought process. The

brain is very similar to most body muscles. It has great adjustability and memory.

When people first join the gym, their muscles hurt and they feel it is the most strenuous thing on the planet. But once their body adjusts to it, muscles start co-operating better and it becomes a lot more enjoyable activity altogether. Also, there is something known as muscle memory. The body muscles get accustomed to the training regime and they grow and function accordingly. This is the physical aspect of the human body.

The mental aspect of humans is not too different from the physical aspect.

Doing mental math is like exercise for the brain. In the beginning it becomes tiring and most people feel the need to quit. But once the mind gets adjusted to the procedures and the step-by-step algorithms, the process gets only simpler with passage of time. It is like walking daily on a tricky road. On the first day it is very difficult. But as you walk on the same path every day, you make marks with your feet and it becomes easier to trace the path as time passes by.

Throughout my childhood, my parents always encouraged me to do math sums mentally. As I grew up it became a habit and I found that mentally I was able to calculate the answers faster than on paper.

In my 8th grade, my dad made me watch a TED talk by Dr Arthur Benjamin, a professor of mathematics at Harvey

Mudd College in California, USA. It was really fascinating to see him do huge multiplication problems in his mind and beat calculators to arrive at the right answers. I was instantly inspired. I was awestruck by how his mind was able to beat all the calculators and I was also highly impressed by the ease with which he presented his show. What followed was a wonderful and memorable story. I practiced some of the basic (not that easy) tricks demonstrated by Dr Benjamin and found myself a platform in my city to demonstrate it. It was difficult at first but I was determined to not give up unless I felt it was totally impossible. My first talk happened and I planned a subsequent eighteen minute talk comprising some elements from Dr Benjamin's show and some of my own. However, at the last moment, the time alloted to me was reduced to a mere eight minutes and in the final talk I had to skip all the elements I planned to borrow from Dr Benjamin's talk. Instead, I ended up doing a bit of music and math and a magic square. It was quite well received and I e-mailed the link of the show to Prof Benjamin and thanked him for being my source of inspiration.

My luck (or destiny rather) played its part and Dr Benjamin responded to my e-mail. It so happened that he was visiting India and he invited me to his talk in New Delhi. Dad and I gladly travelled to the capital city with the hope of talking to this amazing math genius from the US. That day turned out to be one of the best days of my life and I got to spend lots of time with the professor. During his talk he even introduced me on the stage as an example to the audience, to show them that they too could do at least some of the mental math problems that he did, if they were willing to try and practice like me.

This incident inspired me to practice even more. Very soon, giving math shows became my passion. I am the proof to all of you that mental math can be done just with a little bit of practice (slightly more than a 'little bit' though). You just need two things: Practice and Passion. The first word can be executed physically; the second word is all about the mind.

'Mental block' plays a huge part in demotivating many people. Keep an open mind and give this chapter a shot; who knows what the result could be?

In this chapter we will try to study some basic techniques that can be used to simplify many math problems. We will be dealing with the simplest of problems, but with a different approach. Let us try and test the limits of our mental strength. We will not go beyond the four basic tools of arithmetic (addition, subtraction, multiplication and division). The ultimate goal will be to do problems inside our heads, faster than doing it on paper (at least to some extent).

This chapter is highly inspired by Dr. Arthur Benjamin's book, "Secrets of Mental Math"[2]. That book inspired me

---

2   Benjamin, Arthur & Shermer, Michael. (2006). Secrets of Mental Math: The Mathemagician's Guide to Lightning Calculation and Amazing Math Tricks.

and taught me many techniques of fast mental math. However, some concepts are slightly different here. Our goal is to first change some of the basic methods that we follow when doing addition, multiplication, subtraction and division and then to apply our new methods to do large problems and carry out and perform impressive tricks. These tricks can be performed live and can be quite entertaining. More importantly, it will help one to solve quickly most day to day problems as well as problems in many competitive exams where time is limited.

I will share with you the methods which I use. I do not mean to say that I am the only one who is correct, nor do I mean to say that this is the only way to do such problems quickly. You might, as you read this book, get better ideas and come up with faster methods to do the exercises listed. That is where the beauty of math lies: multiple ways to arrive at the same answer and all the methods always work. Just do what works best for you.

Addition

Let us first look at something that is used most often, normally on a day to day basis: simple addition. When solving addition on paper we follow methods which require us to add digit by digit, right to left, with some carry over if required, and then we get the final answer.

The faster and more effective method is to apply the place value system for addition. First, we learn to split numbers into their component multiples of ten. For example, a number like 345 is separated into 300 + 40 + 5, and a number like 4531 is split as 4000 + 500 + 30 + 1.
Thus the first thing to practice is to quickly split number into thousands, hundreds, tens and ones. Try to split the following numbers mentally in less than ten seconds.

Practice 1.1
a. 456
b. 4321
c. 6744
d. 5440

Answers
a. 400 + 50 + 6
b. 4000 + 300 + 20 + 1
c. 6000 + 700 + 40 + 4
d. 5000 + 400 + 40

When we speak about mental math we are literally talking about 'no paper' math. So as far as possible do not use paper even to write down the practice problems. Just look at them and mentally calculate the answers. The goal is not to get the right or wrong answer. At this point, the goal is to train the mind to **think numbers, and think them quick**.

When adding two numbers, only one of those two numbers has to be split. Suppose we are adding 28 and 54, only one of them has to be split (you get to choose).

Let us say we choose to split 54 (50 + 4). Thus, now we have a new addition problem; i.e, 28 + 50 + 4 (instead of 28 + 54). Adding 28 + 50 is as simple as adding 2 and 5. So, 2 + 5 is 7 and 28 + 50 is 78. Now all that is left to be done is 78 + 4 which is 82. Therefore, 28 + 54 is 82. This technique is a lot easier than doing it on paper.

When adding two 3-digit numbers, the method remains the same. For example, 321 + 435. Suppose we choose to split 435. 435 is 400 + 30 + 5. Now, all that remains is to add 321 + 400 + 30 + 5. Here we are playing with the place value system. So all we have to do is carry out a series of 1-digit additions. We first add 321 + 400 by adding the multiples of 100: 3 + 4, which is 7. so 321 + 400 becomes 721; next when adding 721 + 30 we look at multiples of 10: 72 + 3, which is 75, so 721 + 30 is 751; and lastly we add 751 + 5, which is 756. And 756 is the correct answer.

Let us try a slightly harder problem: 876 + 684 (Here we have to have a slightly larger storage capacity in our brains).

Suppose we choose to split 684 (600+80+4), we now have to do 876 + 600 + 80 + 4.

a. 876 + 600 is nothing but 8 + 6, which is 14 and the remaining 76; i.e, 1476

b. 1476 + 80 is 147 + 8, which is 155 and the remaining 6; 1556

c. Lastly, 1556 + 4 is 1560.

This uses a little more memory in the brain, which can be acquired through practice. But, at no point are we dealing with any large addition problem.

This procedure has many advantages.

1. We do not have to worry about carry overs and borrowing (in case of subtraction).

2. Every step takes us closer towards the right answer and we do not have to look back as we follow the algorithm.

3. At no point are we adding large numbers; rather we are dealing with simple 1- digit additions.

Now try these practice problems:

Practice 1.2

a. 435 + 123

b. 532 + 43

c. 567 + 666

d. 111 + 889 (splitting 111 makes it easier)

Answers

a. 558

b. 575

c. 1233

d. 1000

As you practice more, it becomes easier to spot which number has to be split to make the additions easier (it comes out of intuition).

Once you master this, the sky is the limit. You can try to add three numbers, or add 5-digit numbers or even try to add many 2-digit numbers one after the other.

Multiplication

Repeated addition is multiplication. If one can add very quickly there is no reason why one should not be able to multiply quickly as well. Multiplication is the most important foundation for mental math and is very handy for performing many useful tricks. There is nothing more impressive than being able to do rapid multiplication, mentally. Multiplication on paper is even longer than addition on paper and is quite confusing. There are different methods taught in schools for multiplication on paper and sometimes a child can get confused or lost among all the different strategies. The easy way out is doing it mentally (sarcasm not intended).

This however has a pre-requisite: You have to be very comfortable with all the multiplication tables upto 10. For instance if a random product is selected (Eg.: 5 times 7, or 8 times 9), you should be able to answer it instantly. If you are not well versed with the tables, I strongly advice you to take a break and return to this chapter only after you master the tables; otherwise it can be quite cumbersome.

Well, assuming you are well equipped with the necessary tools, let us proceed.
Firstly, we will be doing **2-digit by 1-digit multiplication** (very basic).

For example : 53 * 5.
Recall the splitting strategy used in addition: 53 becomes 50 + 3.
So now the problem becomes (50 + 3) * 5.
Now we apply simple distributive law: (50 + 3) * 5 = (50 * 5) + (3 * 5)
This is nothing but simple 1-digit multiplications.
50 * 5 is nothing but 5 * 5 with a 0; 250.
3 * 5 is 15.
Add the two together: (250 + 15), using the addition methods, to get 265.
Hence, 53 * 5 = 265.

Let us try a slightly tougher example: 79 * 8.
=> (70 + 9) * 8
=> (70 * 8) +(9 * 8)
=> (56 and zero) + 72
=> 560 + 72
=> 632 (using addition technique)

The most important thing is to break the problem down into simpler parts. Since each of the parts can be easily solved individually and mentally, it only requires a bit of practice and brain memory to do them one after the other and then combine all of them .

Try these problems and make sure you do not use paper to solve them. Just look at the problem, memorize it, close your eyes and try to follow the steps discussed above. Write down only the answer so that you can verify all the five problems together at the end.

Practice 2.1
a. 56 * 6
b. 62 * 9
c. 33 * 6
d. 19 * 7
e. 97 * 4

Answers
a. 336
b. 558
c. 198
d. 133
e. 388

2 by 2 multiplication (attempt this only after mastering 2 by 1 multiplication)

If you are able to do 2 by 1 multiplication in less than ten seconds (which I believe is perfectly possible with practice), there should be no difficulty in doing 2 by 2 multiplication.

A 2 by 2 multiplication problem is just an extension of a 2 by 1 multiplication problem. It is similar to doing two seperate 2 by 1 problems and remembering both of them.

Let us start with a simple problem : **23 * 47**.

You might recollect from the addition problems that when there are two numbers, we just split one of them. In this case let us split 23 as 20 + 3.

Thus, the problem now becomes (20 + 3) * 47.

=> (20 * 47) + (3 * 47)

These are two 2 by 1 problems (the first one has a zero but it is just a 2 by 1 problem in disguise)

The first one is 2 * 47 with a '0' = 940

The second one is 3 * 47 = 141 (Following the 2 by 1 technique of multiplication)

The only additional thing to be done (pun intended) is to add the two answers.

=> 940 + 141 is 1081 (Addition method)

We arrive at our answer, 1081.

(Never forget however, that the whole thing has to be done mentally. Hence, unless you can do the previous exercises effortlessly in the head, it is very difficult to even think of these larger problems.)

This might look long. Do not forget, however, that this looks long *only* on paper. Since you are 'reading' these procedures, they might look long. If you are 'thinking' these procedures, the thinking happens very quick.

If you do not believe me, try this example. Imagine you have to remember the name of your dad's brother's wife's brother's daughter. It looks long on paper, but in reality it is just your aunt's brother's daughter. These mental math problems work in a similar way. The more you do them, the more you get used to them and the mind will only work faster and more efficiently with practice and time.

Now try these practice problems:

Practice 2.2
a. 47 * 32
b. 65 * 31
c. 97 * 29
d. 56 * 62
e. 13 * 42

Answers
a. 1504
b. 2015
c. 2813
d. 3472
e. 546

2-digit squares

One of the easiest and most impressive things to be done by mental math is the squaring of numbers. Squaring a number is multiplying the number by itself. For example, 4 squared is 16, 3 squared is 9.

There are two ways of squaring numbers.

1. $(a + b)^2 = a^2 + b^2 + 2ab$
2. $a^2 = (a + z)(a - z) + z^2$

Both the methods can be used to square numbers. To square a 2-digit number, there is absolute necessity of 2 by 1 multiplication as well as lightning quick addition (if you use the second method). Personally I prefer the second method for 2-digit and 3-digit squaring but for numbers beyond that (4-digit and 5- digit squares) I use the first method.

Let me briefly tell you how both the methods work.

Method 1: A 2-digit number, for example, 27 can be split into $20 + 7$

27^2 can be calculated as $(20 + 7)^2$.

$(20 + 7)^2$ is expanded as $20^2 + 7^2 + (2 * 20 * 7)$.

This roughly translates into 2 by 2 multiplication, and a few simple multiplications and then an addition of all the three.

Method 2 : Multiplying a 2-digit number by itself might be hard. So we pick a number near to the 2-digit number that is easy to multiply by.

For example, 23 * 23 is hard, so we try to multiply by 20 instead. We pick 20 as it is near 23; Likewise, 46 * 46 is hard so we use 50 instead (50 is near 46).

As you might have noticed, we pick an easier number (lesser than or greater than our initial number) to multiply by. We have to maintain an equilibrium, however.

Take 23 for instance. If we are using 20 instead of a 23, we would also have to use a number slightly larger than 23 (since 20 is smaller than 23). We thus have to go up or down equally on either side of the original number. Thus 23 * 23 can be changed to 20 * 26 (lot easier to handle because of the zero). And likewise, 46 * 46 can be changed to 42 * 50 (notice how we have maintained equal distance on either side of the initial number, hence maintaining the equilibrium).

In short, we are converting the same problem into a simpler problem by substituting numbers nearby.

Let us look at the example again and go ahead with the remaining steps.

23 * 23 => 20 * 26 = 520

The final step is to add the square of the number which is the difference between the initial number and the new numbers (in this case: 20 is 23 − 3 and 26 is 20 + 3; thus the difference number is 3). Hence we add 9 to get the final answer : 529.

Let us now look at the second example:

46 * 46. We use 50, which is 46+4; or going up by 4.

So to maintain equilibrium, we need to go down by 4 as well.

=> (46-4) * (46+4) + (4 * 4) {because 4 is the difference number}

=> 42 * 50 + (4 * 4)

=> 2100 + 16

{42 * 5 is done by the 2 by 1 technique with the addition of a zero at the end}

The easy number could be smaller or larger than the original number. The important thing is to maintain the equilibrium.

Now try these practice problems:

Practice 2.3

a. 47^2

b. 67^2

c. 72^2

d. 54^2

e. 97^2

Answers

a. 2209

b. 4489

c. 5184

d. 2916

e. 9409

3 by 1 multiplication and 3-digit squares

To perform 3 by 1 multiplication and 3-digit squares, we use the same logic used for 2-digit multiplication and squaring. The splitting strategy of numbers remains the same.

3-digit squaring can be done by first practicing 3 by 1 multiplications. The principle remains the same : 347 * 8 can be written as $(300 + 40 + 7) * 8 = (300 * 8) + (40 * 8) + (7 * 8) = 2400 + 320 + 56 = 2776$.

Practice 2.4

a. 435 * 4
b. 324 * 7
c. 521 * 9
d. 319 * 4
e. 913 * 6

Answers

a. 1740
b. 2268
c. 4689
d. 1276
e. 5478

The principle remains the same for 3-digit squaring (same as the method to do 2-digit squares).

Let us take a number, say 378. At first look, one might feel that 380 is the number which is easy to multiply by. However, there is an easier number...400. So here we look for an easier number closer in the hundreds rather than just picking numbers closer in the tens.

So, if we replace 378 * 378 by 356 * 400 (equilibrium has to maintained so we have to increase and decrease equally by 22, since 400 is 22 greater than 378), what are we left with? Well, the square of the difference, which in this case is 22 * 22 .

=> (356 * 400) + (22*22)

=> 142,400 + 484 (3 by 1 multiplication and 2-digit square)

=> 142,884.

3-digit squares are a lot harder than 2-digit squares and can be done only after being very comfortable with 2-digit squares, 2 by 1 multiplication and 3 by 1 multiplication. Another example : 742 * 742. This can be written as (700 * 784) + (42 * 42).

=> 548,800 + 1764

=> 550,564 (the addition has to be done by splitting 1764 as 1000 + 700 + 60 +4)

Try these problems.

Practice 2.5

a. 542^2

b. 789^2

c. 143^2

d. 388^2

e. 965^2

Answers
a. 293,764
b. 622,521
c. 20,449
d. 150,544
e. 931,225

Calculating the day of the week for a given date

Let us take a break from multiplication and learn a few math tricks. This is one of the most impressive things I ever saw. Imagine, being able to determine correctly the day of the week for any given date (past, present and future)! I first thought it was only possible through supernatural means (some intuitive guessing), but it turns out there is a perfectly logical mathematical algorithm to do this with precision for literally any date.

Follow the steps given below and track the example that we are using:
1. First get the year (Let us suppose that the year is 1977) Take the last two digits of the year
(eg: 77 of 1977)
Divide it by 4. Keep the quotient and ignore the remainder.

Add this to the original 2-digit number. (Eg: 77/4 gives us 19, 19 + 77 = 96)

Add 1 to the answer if it is 1900s. Add 3 if it is 1800s. Add 5 if it is 1700s. Add 0 for the 2000s. (Eg. : Since it is the 1900s we add 1 and we get 97)

Divide this by 7. Forget the quotient. Keep the remainder. Let this be X

(Eg: 97/7 gives us remainder as 6. So, X = 6 in our example)

2. Next get the month. Add the respective month code (Each month has a specific digit associated with it)

Jan	Feb	Mar	Apr
6	2	2	5
May	Jun	Jul	Aug
0	3	5	1
Sep	Oct	Nov	Dec
4	6	2	4

(Let us assume that the month is March. So we add 2 to our previous answer of 6 to get 8)

Note: If it's a leap year, Jan code is 5 and Feb code is 1

Add the code to X. Let this be Y (6 + 2 = 8 = Y in our example)

3. Finally, get the date.

Add the date to Y. Let this be Z (Let us assume that the date is 15. So adding 15 to Y we get 23)
Divide Z by 7. Keep the remainder (23/7 gives remainder 2)

Follow this code

Mo	Tu	We	Th	Fr	Sa	Su
1	2	3	4	5	6	0

Hence, the final remainder corresponds to the final answer; i.e, the day of the week. (Hence, in our example, we arrive at Tuesday, corresponding to the remainder 2. Thus, 15th March, 1977 was a Tuesday.)

This process is probably the hardest to memorize but since every step involves hardly anything tougher than basic arithmetic, it can become trivial after proper practice.

Practice 3.1
a. 1942, March 13
b. 1947, August 15
c. 2000, April 6
d. 1970, July 17
e. 1854, October 10

(For leap years, the month code for Jan and Feb changes; everything else remains the same. Keep in mind though, that 2000 is not a leap year.)

Answers
a. Friday
b. Friday
c. Thursday
d. Friday
e. Tuesday

This same formula can be used to calculate any day of the week from 1600 to 3000. You might recall that during the calculations we have to add 1 if its 1900s, 0 if its 2000s, 3 if its 1800s and 5 if its 1700s. This addition part repeats every four hundred years. Hence, 1600, 2000, 2400, etc. have the same addition part (adding zero). The same applies for 1700, 2100, 2500, etc. (adding 5).

Some interesting trivia: September and December have the same calendar every year; January and October have the same calendar for non-leap years; April and July have the same calendar every year and so do March and November.

Birthday Magic Square

Srinivasan Ramanujan, one of the greatest ever mathematicians to walk on planet Earth was very famous for his plethora of theorems, proofs and conjectures. However, I got to know about him through something very fundamental and simple, yet fascinating.

During my 8th grade, my dad sent me a magic square done by Ramanujan which was designed in a very unique way. It had his birthday in the top row and it had multiple ways of arriving at the magic sum (not just rows, columns or diagonals). I tried to make one for my own birthday, and by sheer fluke ended up doing it late one evening. At that time I did not follow any particular algorithm to construct the birthday magic square; so I really did not know how I had done it.

Over time, as I began giving shows and doing magic squares live in front of people, I was able to come up with a method of actually constructing it.

Before I share with you the method, let me first show you the Ramanujan magic square.

22	12	18	87
88	17	9	25
10	24	89	16
19	86	23	11

22nd December 1887 is Ramanjun's birthday and this particular magic square has about 30 different ways in which the same answer of 139 can be obtained.

Here is a magic square which I have created for my birthday 13th March 1999:

13	3	19	99
94	24	14	2
11	5	91	27
16	102	10	6

If you observe the two magic squares (Ramanujan's birthday square versus mine) you will not notice any obvious similarities. I have absolutely no idea what was Ramanujan's strategy. My method might not be the only method of doing this wonderful box, but I can assure you that it works most of the times.

One of the major restrictions in the magic square is to not repeat any numbers. The beauty lies in creating a birthday magic square with sixteen unique numbers and having multiple possible ways of arriving at the magic sum (which in my birthday square is 134).

Now I will explain the method of constructing the birthday magic square:

Let the numbers in the birth date be A B C D. So, for 13 March(03), 1999, A is 13, B is 03, C is 19 and D is 99.
Let the numbers in the magic square be filled from A to P as follows:

A	B	C	D
E	F	G	H
I	J	K	L
M	N	O	P

Choose a small number to your convenience. It could be 1, 2, 3 etc. Let us name this X.

Subtract X from A and get O. O = A + X

Add X to B and get P. P = B - X

Subtract X from C and get M. M = C - X

Add X to D to get N. N = D + X

Obtain J and H such that J − H is equal to X.

Rest of the numbers are created automatically by filling every row, column or diagonal such that they add up to the magic sum. Magic sum is A+B+C+D.

If any of the numbers repeat, then it means that the choice of X is wrong.

Change X and repeat the above procedure. Or you may also keep the same X and change J and H instead. This might seem like trial and error, but with practice, it is possible to get ta 'feel' for X, J and H such that they create a box without repetition. However, keeping the repetition criteria aside, this algorithm ensures the following :

1. All the numbers in each row add up to the same magic sum, S.

2. All the numbers in each column add up to S.

3. All the numbers in each diagonal add up to S.

4. Multiple other combinations (such as numbers in the four corners; four numbers in the center; numbers in any 2 by 2 boxes within the 4 by 4 square; etc.) all add up to the same magic sum, S.

I will not explicitly state all the combinations of arriving at the magic sum. There are more than thirty of them. Try for yourself and see how many you can spot!

Here is an 8 by 8 magic square that I created for my birthday (13 3 19 99):

1300	300	1900	9900	1287	313	1887	9913
8800	3000	1600	0	8787	3013	1587	13
1500	100	8700	3100	1539	113	8687	3061
1800	10000	1200	400	1787	9961	1239	413
1270	330	1870	9930	1201	399	1801	9999
8770	3030	1570	30	8701	3099	1501	99
1590	130	8670	3010	1797	199	8601	2803
1770	9910	1290	430	1701	9703	1497	499

I have used 4-digit numbers so that there will be no repetition (there are sixty four boxes to fill, hence using 3-digit or t2-digit numbers increases the chance of repetition. This magic square was created when I was in my 9th grade. You might wonder what is so special about this box. Well, it exhibits all the basic magic square properties (rows, columns, diagonals: each adds up to the same magic sum 26,800). In addition, it has lot of other magical properties as well. The magic sum here is 26,800. Thus, any which way you add, 26,800 is the answer.

So, how many 'any which ways' are there? More than three thousand! Yes! The birthday magic square had about thirty six ways of arriving at the magic sum; here it is more than three thousand! Before I tell you how to construct this 8 by 8 beauty, let me show you some of the magical patterns which add up to the magic sum (in addition to the rows, columns and diagonals).

1300	300	1900	9900	1287	313	1887	9913
8800	3000	1600	0	8787	3013	1587	13
1500	100	8700	3100	1539	113	8687	3061
1800	10000	1200	400	1787	9961	1239	413
1270	330	1870	9930	1201	399	1801	9999
8770	3030	1570	30	8701	3099	1501	99
1590	130	8670	3010	1797	199	8601	2803
1770	9910	1290	430	1701	9703	1497	499

These might look like diagonals but they are actually different types of diagonals. They are formed when one diagonal is cut in half and continued in another direction. Thus, this magic square makes it possible for different pieces of diagonals to still merge and add up to the magic sum of 26,800 in addition to the normal diagonals doing the same. We will not go into each of those three thousand plus combinations; hence I will allow you to trust me and believe it!

1300	300	1900	9900	1287	313	1887	9913
8800	3000	1600	0	8787	3013	1587	13
1500	100	8700	3100	1539	113	8687	3061
1800	10000	1200	400	1787	9961	1239	413
1270	330	1870	9930	1201	399	1801	9999
8770	3030	1570	30	8701	3099	1501	99
1590	130	8670	3010	1797	199	8601	2803
1770	9910	1290	430	1701	9703	1497	499

What is the secret behind the 8 by 8 box ?

The secret lies in each of its four quarters.
If you notice closely, each of the quarters, the smaller 4 by 4 boxes, is a magic square in itself.
The amazing thing is that each of these 4 by 4 magic squares has the same magic sum of 13,400.
This implies that the 8 by 8 box is actually a combination of four 4 by 4 magic squares, each having the magic sum of 13,400.

1300	300	1900	9900	1287	313	1887	9913
8800	3000	1600	0	8787	3013	1587	13
1500	100	8700	3100	1539	113	8687	3061
1800	10000	1200	400	1787	9961	1239	413
1270	330	1870	9930	1201	399	1801	9999
8770	3030	1570	30	8701	3099	1501	99
1590	130	8670	3010	1797	199	8601	2803
1770	9910	1290	430	1701	9703	1497	499

The even more magical thing is that, not only are these 4 by 4 boxes magic squares, but they are also special magic squares with about twenty-four ways of arriving at the same magic sum of 13,400 (in each of the smaller boxes, rows, columns, diagonals, center numbers, corner numbers and various other symmetrical patterns total up to the same sum, 13,400).

Therefore, we can pick one row from one quarter (small magic square) and add it to one row or column or diagonal from any of the other three quarter magic squares and get a sum of 13,400 + 13,400, i.e., 26,800. Since there are twenty four possible ways of arriving at the magic sum in each of the quarters, and there are four such quarters, the total permutations add up to more than three thousand!

A note on subtraction:

All of the multiplication problems attempted by splitting of numbers (27 as 20 + 7) can also be done by the subtraction method. 29 can be written as 20 + 9 but the easier way to write it could also be 30 – 1.

So when multiplying a number by 29, for example : 46 * 29, it can also be written as 46 * (30 – 1) which then becomes (46 * 30) – (46 * 1).
Doing subtraction mentally is very similar to doing addition. It has to be broken down into simpler parts and the rest just follows the same principles.

For example :
366 – 231 = 366 - (200 + 30 + 1) = 366 - 200 - 30 – 1 = 166 – 30 – 1
= 136 – 1 = 135

Try this examples
Practice:
    1. 452 – 311
    2. 8992 – 2314
    3. 4228 – 311

4. 2299 – 421
5. 8732 – 5219

Answers:
   1. 141
   2. 6678
   3. 3917
   4. 1878
   5. 3513

The entire theory is all about learning how to do simple arithmetic mentally and learning how to break a huge problem into simpler parts and doing all steps without forgetting the previous calculations.

The question then arises, 'how is it possible to do this if someone has never done it before?'. Human memory has a remarkable characteristic which may be termed 'memory stretchability'. One might wonder about one's ability to remember so many events as one's life goes on. How does one not forget what happened in one's childhood or school days? The human memory stretches as time passes. It has infinite capacity. It is not like a mobile phone which has a limited storage space restriction. Human memory has the ability to expand and stretch its storage capacity over time. When one first attempts mental math it is understandably hard. But all it requires is patience. A simple way to test the stretchable power of the human mind is to learn a new word every day. On the first day, one learns a word. The second day, one should recite the word learnt on the first day as well as the new word learnt on that (second) day. So each day, one should recite all the words learnt on previous days and the new word learnt that day. This process can be expanded for a really long time and the mind will never run out of space. This also happens when we see new faces, memorize new mobile numbers, make new friends, read new books and watch new films. The information just keeps adding up and the mind never runs out of room. The only thing to be done is to have a fresh mind and a completely clean brain RAM (especially to do 3-digit squares and above). Thus, do not try to do any mental math calculations as a parallel activity or as part of a multi-tasking activity session.

There is an online game 'Tux of math command'3 (TuxMath for short), which requires the player to type answers to single variable equations in less than few seconds. I would play that game during the wee hours of the mornings when in 8th grade. When you play a game, your mind will be fooled into thinking that it is not studying. It makes you think that you are actually doing a fun activity (if studying can also be a fun activity, nothing better). So, as you eventually get better at the game as you try to win and beat your previous scores, your mental math speed as well as accuracy improves significantly.

---

3   https://en.wikipedia.org/wiki/Tux,_of_Math_Command

So far, we have briefly looked at various types of mathematics that is very impressive to be presented to an audience, helpful for development of ability to do math problems mentally and also useful to increase the overall capacity of the mind to handle problems. If you find this aspect of math very interesting, I strongly recommend Dr Arthur Benjamin's book 'The Secrets of Mental Math'. Note that we have not done anything from the widely known: Vedic Mathematics of Swami Bharati Krishna Tirtha. What I have explained in this chapter is all basic level of math that does not involve any tricks, but involves practice and application of simple math logic and formulas to help us do the problems mentally, and quickly. I strongly believe that there is no 'short cut' methods to do mental math and the only way to do it is to break the problem down into simpler parts and follow the basic, fundamental rules of arithmetic.

Many books talk of faster methods for multiplying, adding and dividing numbers. But more often than not, these methods are restricted to only certain types of numbers (many of you might be aware of the quicker ways of squaring numbers that end in '5') and are not universally applicable to numbers as a whole. Thus, I find it safer and faster to use normal algebraic techniques rather than memorize a number of short cut methods and then lose time trying to figure out which short cut to apply where. Hope this makes sense!

5. WHERE DO WE SEE MATHEMATICS?

Look for something with an open mind,
not just an open eye.

- Me

We have discussed a lot about mathematics and how it originated in ancient India. Most people who initially found math difficult, I hope, have given it a second thought. Those who found it difficult to do mental math, I hope, have tried to learn it and approach it again with an open mind.

Pre-historic man was unaware about the intricacies of math and for him it was all about counting and keeping track of a few day to day objects. For him, numbers were the basic, necessary methods for keeping tally and giving quantitative value for various topics. As time progressed, the subject took a broader meaning in society and math became the mother from which various ideas took birth. The medieval period saw the rise of many math topics such as algebra and trigonometry.

Ancient Indian mathematicians found applications for mathematics in their everyday life and they had their beliefs and traditions for which mathematics acted as a fool-proof tool. They operated independent of the rest of the world and had unique approaches for popular concepts.

In today's busy world, mathematics has become nothing more than one of six major subjects taught at school (at least for the typical adolescent). It has become a fear for many, and most people feel it to be redundant. Many people feel math as something 'not necessary' and 'not useful'. As you might recall from the first chapter, math is a language; and a very subtle one at that. If you know where to look, and how to look, you will most definitely conclude that it is the language in which our entire universe has been written. For starters, there is a branch of mathematics that deals with the concept of chance and randomness. We call it 'probability'. The concept of probability is vital in studying the existence of the universe. What is even more interesting is that probability can be used to prove or disprove many beliefs and myths; for example, the existence of God. Most theists (believers of God) use probability to prove to the world that God does exist.

It so happens that the probability that the universe exists is infinitesimally small. So tiny that the probability that the universe exists and expands the way it is expanding now is approximately similiar to the probability of rolling a dice seventy-two times consecutively and arriving at the number 'six' every single time! Since this is almost impossible to happen by chance, many great thinkers feel that there is a creative genius who is responsible for the creation of this wonderful world; and they refer to the creative genius as 'God'.

Probability is wonderful. It is fun. It is virtual; however it is applicable in the real world. Sports, gambling, world statistics are all formulated on the principles of probability. Probability is ultimately a branch of mathematics; hence without mathematics it would be almost impossible to come up with the concepts of 'prediction', 'analysis' or 'randomness'.

Logic is the foundation on which mathematics is built. Math is the only subject that is constructed entirely on logic, facts and proofs and with very little assumptions. The beauty of math is that it is always accurate. It is reliable. In its own way, it is always correct. It is the invisible robot that obeys and performs all commands without fail. Mathematics works on the principle of logic. Hence, there are very few illusions and very few lies. However, since it is always right, no matter what, illusions or lies created with the help of mathematics and logic, will always be more effective. Thus, we arrive at the king of illusions, Magic.

When I say Magic, I am not referring to the 'rabbit out of the hat' or the 'bird coming out from the sleeve'. Magic created with the help of logic and mathematics is truly, *Magical*. The birthday magic square serves as a working example of mathematical magic. It is a magic trick that can be taught, explained, performed with close up camera shots, and still become more impressive and mind–blowing than many a magic trick. Mathematics is mysterious and although applicable in many fields, its beauty lies in its invisibility and virtual nature. You cannot actually see math, but you know it exists. You have to prove it or disprove it, without any actual evidence of its existence in the real world. Math is all about patterns; it is a wonderful, flawless design; it is never ending and it has no definite beginning.

The most mysterious of mathematical concepts are the prime numbers. They have kept mathematicians interested from many centuries and yet fail to lose their charm. There is always scope for invention and there is always room for development. The fact that they do not have any factors, makes them very difficult to locate. Two prime numbers when multiplied yield a new number, their product, which has only those two prime numbers, and no others, as its factors. This makes the product highly important in fields such as cryptography. Since no one can ever know those prime factors (if they are large enough), such factors can be used to develop codes or secret languages to hide or communicate numbers that are highly confidential (for example, banks use prime numbers to encrypt and communicate confidential financial information).

Mathematics is something that very few people actually understand (no one can ever understand it completely). This makes it invulnerable. If someone studies prime numbers deeply and develops a cryptic code using them, it is almost impossible for anyone to crack it except the one who developed it. Mathematics is highly secure, well-defined and developed.

Numbers give a quantitative sense to the world. Without numbers it is difficult to compare. Numbers are a virtual way to define 'more', 'less', 'a lot' and 'nothing'. Data can be understood and handled with ease when given numerical values. Topics can be visualized with clarity when explained quantitatively. Most importantly, numbers help in creating concepts like 'precision' and 'estimation' in the human mind. Our sense of quantity and our understanding of logic is what potentially makes us different from most beings living on this planet.

Mathematics is all about patterns. There are rules for the behaviour of even the tiniest of math concepts. Math gives so much importance to precision, that even the slightest of mistakes makes the answer completely wrong. There is no credibility or consolation for 'partly correct' or 'almost correct'. It is par excellence and a stickler for perfection. This connects it to something else which too follows all of the above statements: the melody which the world is blessed to have, viz., music.

Music works in a weird way. When performed perfectly, it gives the ultimate satisfaction. However, the slightest of errors makes it horrible and unforgivable, very much like mathematics. Hence, the two go hand in hand.

As you might recall, in ancient India, people were multi-talented. More often than not, a mathematician would apply his sense of numbers for the benefit and flourishment of his other talents such as music.

One such mathematician was Acharya Virahanka (6th century AD). Since his original ideas were applicable to Sanskrit poetry, we will not dwell too much into that. However, Prof Manjul Bhargava of Princeton (one of my greatest inspirations) wonderfully explains this concept on the Tabla[4] (he, like the ancient Indian mathematicians is a multi-talented genius). Being a tabla player myself, I found this very exciting as well.

Let us try and understand how math can actually help with music.

In tabla, there are two types of syllables (just like in Sanskrit) : Long syllables and short syllables (recall Pingala). Here, however, we represent the short syllables with '1' and the long syllables with '2' (as it is not possible to play or represent '0' on the tabla, as you have to play 'nothing').

The short syllable (1) can be called 'Laghu'.

The long syllable (2) can be called 'Guru'.

Virahanka was a poet; so he wanted to compose poems such that they rhyme and have rhythm. If a song has to be rhythmic, each line of the song should take the same amount of time when sung or played.

Hence, we have two tools (Laghus and Gurus) to construct poems or songs or tabla beats.

Let us assume that each line of our song has a time duration of '4' units.

This gives us the following five combinations of Laghus (1) and Gurus (2):

$1 + 1 + 1 + 1$

$1 + 1 + 2$

$1 + 2 + 1$

$2 + 1 + 1$

---

4   https://www.ams.org/programs/students/wwtbam/arl2006

2 + 2

which add up to our required sum of '4'.

We, thus, have five ways of adding up to '4' using '1' and '2'.

Let us try the same exercise for the number '5' (assume each line of the song has a time duration of '5' units)

We get the following combinations of Laghus (1) and Gurus (2):

1 + 1 + 1 + 1 + 1
1 + 1 + 1 + 2
1 + 1 + 2 + 1
1 + 2 + 1 + 1
2 + 1 + 1 + 1
1 + 2 + 2
2 + 1 + 2
2 + 2 + 1

That gives us a total of eight ways of arriving at '5' using '1' and '2'.

If one takes the trouble of actually listing all the combinations of '1' and '2' to arrive at various time duration units, we get the following results:

Time duration	0	1	2	3	4	5	6	7	8
Combinations	1	1	2	3	5	8	13	21	34

The above table gives us the number of possible ways of arriving at a number as a sum of '1s' and '2s'. If you notice the 'combinations' row carefully, it gives the Fibonacci sequence as a result.

(Note: to arrive at zero, there is only one way and that is to not use '1' or '2' in the combinations. 'Doing nothing is still a way of doing something')

Enumerating the various combinations of '1' and '2' can be really useful in music as it gives the composer multiple combinations of notes to compose songs. Each line can be one of the many unique combinations and the wonderful thing is the total sum remains the same (thereby ensuring that the line will maintain the rhythm.

Since music is all about having various combinations of numbers adding up to the same sum (to maintain rhythm), a wonderful mathematical tool for music would be the magic square. A four by four magic square (when created in the way discussed previously) has more than thirty ways of adding up to the same answer (magic sum). Corelating this to music would certainly be something special.

Violin is the other musical instrument that I play (in addition to the tabla). I play Indian classical music on the violin; it basically consists of eight notes:

Sa Re Ga Ma Pa Da Ni Sa (same as **Do Re Mi Fa So La Ti Do)**.

These eight notes can be played or sung in many different combinations to create melody. However, we need to maintain rhythm. Hence, each combination should take the same amount of time. For most songs, the time duration of 16 units is melodious (powers of 2 somehow create the intuitive rhythm sense to which most people respond. This is the 'catchy' beat). So let us create a 4 by 4 magic square that adds up to sixteen and let us use numbers ranging from '0' to '8'.

4	4	4	4
2	6	8	0
7	1	1	7
3	5	3	5

Thus we now have a magic square which has a magic sum of 16 and has more than thirty ways of arriving at it through various ways (rows, columns, etc.).

Instead of looking at these numbers as mere digits, let us convert them into musical notes.

0 refers to playing 'nothing'.

1 is 'Sa' or 'Do', played for 1 unit of time.

2 is 'Sa Re' or 'Do Re', played over 2 units of time.

3 is 'Sa Re Ga' or 'Do Re Mi', played over 3 units of time.

....

....

....

8 is 'Sa Re Ga Ma Pa Da Ni Sa' or 'Do Re Mi Fa So La Ti Do', played over 8 units.

Therfore, if we want to play the rows of the magic square on the violin, we can play them as:

1st row : 4 4 4 4 : 'Sa Re Ga Ma' + 'Sa Re Ga Ma' + 'Sa Re Ga Ma' + 'Sa Re Ga Ma'

2nd row : 2 6 8 0 : 'Sa Re' + 'Sa Re Ga Ma Pa Da' + 'Sa Re Ga Ma Pa Da Ni Sa'

3rd row : 7 1 1 7 : 'Sa Re Ga Ma Pa Da Ni' + 'Sa' + 'Sa' + 'Sa Re Ga Ma Pa Da Ni'

4th row : 3 5 3 5 : 'Sa Re Ga' + 'Sa Re Ga Ma Pa' + 'Sa Re Ga' + 'Sa Re Ga Ma Pa'.

Use the same approach for columns, diagonals, corners, center numbers and all the other possible combinations of arriving at the magic sum of 16. Thus, one simple magic square gives us multiple ways of playing a rhythmic melody.

Note (For the music enthusiasts): In Indian classical music there is something called 'Swara Prasthara' in Carnatic classical, or 'Taan' in Hindustani classical, where the artist has to compose combinations on the spot. These combinations have to be mathematically accurate (should always end at the 'sam' or starting beat) as well as musically accurate (should not stray away from the Raaga). This idea of magic square gives ready made permutations for 'Swara Prasthara' or 'Taan' and the artist can plug in a different magic sum based on the 'Taal' (rhythm cycle) and the notes to be played can be decided based on the Raaga (I have explained using the eight note 'Shankarabharanam' to make it appealing to the neutral audience who might not be aware of the other Raagas).

Hence, we realize that the principles of math are applicable in many areas. I was initially interested in astronomy and cricket. Later I realized that what I really liked about astronomy was the comparison of star sizes, studying their temperatures and estimating their volumes with respect to the sun and the earth. I realized that what I really liked in cricket was statistical analysis, memorizing averages and scorecards and also comparing batsmen and bowlers by using various statistical techniques (Eg., using arithmetic versus geometric mean to compute performance).

Thus it was the numerical part of cricket and astronomy that fascinated me. It was actually the math behind those topics that appealed to me. Math has had lots of influence on almost every science in the world because without statistics, quantitative analysis and logical thinking, it is difficult to see any progress or growth in any area of interest.

6. MATHEMATICS IN STORIES

Everything can be explained through stories.
-Me

This chapter consists of many math problems and puzzles (no riddles) that I have gathered and admired throughout my life. Some of these are borrowed from books and people and some of them are created by me. The answers are at the end. Please do not look at them until you are done with all the problems. Each of these problems will be worth certain amount of points (based on the difficulty level). Please grade yourself and feel happy irrespective of whether you do badly or pass out with flying colours, because if you can even read and get through all the problems, I would have indeed succeeded in planting the seeds of mathematical and logical interest in you.

Although some of these problems are borrowed, the stories and poems are completely original. Enjoy !

1. The Dilemma of the Traveler

Prem was really delighted. He had won the lottery. He did not exactly win as much as he had expected but what he had won was definitely sufficient for him to fulfill his dream of visiting his dream city.

His work commitments had perfectly aligned to facilitate his trip. He was on a much needed vacation. He was on the final flight that would lead him to his destination. He looked out of the window and pondered over this amazing place that he was about to visit.

Located in the interiors of Assam, it was not very well known to most of the world or even the country. Prem had read about it and with the help of his uncle living in Assam, he had investigated the paths that would lead him there. The city was supposed to be amongst the top most beautiful places on planet Earth. It had the least pollution, the best green trees and the most unbelievable cuisine. However, the main attraction was the nature of the residents: they all spoke only the truth!

The city was estimated to have about a thousand residents and every single one of those spoke only the truth. The entire city was infact called 'Truth city' since each and every person there just spoke the truth, the complete truth and nothing but the truth. For example, if a couple went out on a date, the girl would instantly say 'I am with you only for your money', or at school a child would say 'I did not do the homework because I am not interested'. The default psychology of the people there was to speak only the truth. Prem was a follower of Raja Harishchandra of ancient India who was very well known for his sincerity and quality of always speaking only the truth. Prem, thus wanted to visit this city as quickly as possible and interact with its citizens.

He got down from the aircraft, booked a cab and was on his way. He reached the spot which was the furthest that the cab would go. He would have to walk four miles from this spot. He was really excited and though not a genuine athlete, the motive of reaching the destination made four miles look hardly anything tiring.

The first two miles went by very quickly. He admired the green trees on both sides of the path that he was treading on. The scenic beauty was something so enjoyable that he did not pause to even click pictures; he decided to just enjoy it (he felt that the pictures could be taken on the way back).

As he walked, he saw a sign board approaching. It was about two hundred yards away and he could not exactly read what it said. As he walked towards it, the writings on the sign board became more legible and he encountered one of the hardest situations of his life. The sign board was at a fork where the path diverged into two. Just below the sign board stood a man of not more than thirty. He was dressed in a wonderful black jumper and looked bored. The sign board read:

Welcome O dear Traveller to the wild,
I do not care if you are man, woman or child.
Two roads diverge in this deadly wood,
Read this message to proceed where you should.

One leads to the city of Truth and Sincerity,
The other leads to the city of Lies and Falsity.
Honest people form the city of Good,
While dishonesty is the policy of the city of Falsehood.

The man could be from either of the two,
Liar or Truth teller, you don't know he is who.
O Traveller, ask him one and just one question,

Based on his answer, decide your destination.

What is the one question that Prem should ask the man to find the right road to the city of Truth?
(3 points)

2. Scaling New Heights

It was a new day in the life of Sameer, the gardener. He lived in the beautiful city of Manipal, India. He looked after a huge garden which was owned by his master. It had been over a month since he had started working. It had been a wonderful experience so far and he was loving every day at work.

He got dressed and visited his master. His master looked amazingly happy and his face was beaming. He was standing holding a newspaper in his hand. Sameer was extremely curious to see the newspaper. On the front page, an article read:

"Competition For Best Garden: on the 13th of March, judges from all over India will visit various cities to decide the best garden in the country. The best garden in every state will receive a cash prize of Rs. 1,00,000 and the best garden of the entire nation will win an astounding sum of Rs. 10,00,000.

Criteria :
There should be exactly one hundred plants in the garden. The average height of the saplings should not be less than two feet.
Each shrub which is larger than 2 feet reduces points. Genetically modified seeds are not allowed. Artificial fertilizers are not allowed."

Sameer was very happy. He promised his master that he would try his best to build the best possible garden. He decided to buy the seeds and get involved with the work as quickly as possible. Managing one hundred plants was going to be quite a lot of work.

As he set about his planning, something from the article struck him *'Average height of the plants should not be less than two feet'*. Thus, he had to grow about a hundred plants and most of them had to be at least two feet or above. He also remembered 'Each shrub which is larger than 2 feet reduces 2 points'. He felt that it was very difficult to grow so many plants that high in less than a month. He thought to himself "I have to try and construct my garden such that there are as many short plants as possible". This way, he would have fewer tall plants and more number of shorter plants. It would reduce the deduction of points because of tall plants.. Thus, he had to construct few plants which are above two feet and plenty of plants that are below the height of two feet.

What is the maximum number of plants that can exist in the garden which are below the height of two feet such that the overall average is still two feet?
(2 points)

3. Soot for those in suits

It was 8 am and the train started off from Mumbai on its journey to the garden city of India, Bangalore. Most of its three hundred passengers were lost in their own worlds but for the three gentlemen sitting in one of the last compartments; they were lost in the world of mathematics. Devilal, Veer and Avinash were three famous mathematicians from Mumbai, Chennai and Kolkata respectively. They had met after a long period of ten years for a conference in Mumbai and were now on their way to Bangalore for an international conference on algebra. They had been fast friends from their first year at college and had many a thing in common, math being the foremost of all. It was a pleasant reunion for men who had excelled in their common field of interest.

The three of them were so engrossed in their conversations that any passer by would be ignored and if not for the occasional stops at stations, they would be completely unaware of their journey or their mode of transport. The discussions shifted from prime numbers to elliptic curves to calculus to trigonometry as the train sped from Dadar to Kalyan to Lonavala to Pune.

It was soon noon. The train traveled at breakneck speed and before the three of them knew it, it approached a long, dark tunnel.

After what seemed like hours, the train came out of the tunnel and there was a surprise in store for the three geniuses. All their faces were covered in soot! As soon as the train journeyed into broad daylight, each mathematician noticed the soot covered faces of the other two and started laughing, unaware that he too was covered in soot.

This continued for about two minutes when Devilal suddenly realized that his face was covered in soot too. There were no mirrors, nor did he smell, touch or feel soot on his face. It was his highly logical mind that made him realize this.

How did Devilal realize that he too was covered in soot? (3 points)

4. Efficient Nephew

Radhe was on his way to his uncle's village. It was his summer vacation and the prospect of playing with his old friends, climbing palm trees and working in his uncle's shop during the day made him very excited.

His uncle was a business man and owned a small general store. Almost anything needed for day-to-day life was available in the store. From comb to medicine, rice to toothbrush, almost every kind of item was found in his uncle's store. Since he had almost nothing to do the entire day, (there was hardly any mobile network in the village), Radhe had decided to work with his uncle and help him manage the shop.

The shop also sold lots of artificial jewellery and text books for students. Radhe was extremely excited as he began work at the shop the next day.
It was only 10 am and customers were already flooding into the shop. His uncle was busy at one end of the shop and Radhe was totally occupied with five customers at the same time; all were shouting at the top of their voices, complaining, bargaining and also quarrelling amongst themselves.

It was soon noon. A tall customer arrived with a pipe in his hand. He had the typical Sherlock Holmes look, thought Radhe. He wanted to buy a couple of books for his son. "Are those real?" he asked pointing to a jar of rings. "Oh no, they are artificial and very inexpensive. They do look very real though," said Radhe, eagerly hoping that the man would buy some as business had not been the best after the loud morning customers' exit.

"How very strange. They look exactly like the ring I am wearing," said the tall customer. "Mine is real though," he added in a light hearted manner. He took out his ring and walked towards the jar of artificial rings. He opened the jar to have a closer look at the rings, when all of a sudden, he sneezed and his ring (which he had held carelessly between his fingers) flew out and landed straight into the jar! The tall man let out an audible gasp. "My ring, my expensive ring!" Radhe and his uncle came to the scene within seconds. The ring was lost in the jar and all the rings looked completely identical.

Radhe thought for a bit. Suddenly, he had an idea. He asked the customer, "How heavy is your ring, Sir?" The man replied instantly, "My ring weighs exactly 3.2 grams and I bought it exactly thirty eight days ago." Radhe then turned to his uncle, "How many rings are there in that jar and do you know the weight of each of those artificial rings?" His uncle thought for a moment and replied, "Each of those artificial rings weighs exactly 3 grams and there are exactly eighty rings inside that jar."

'Get me a pair of scales, I don't need any weights," said Radhe to his uncle. The customer was shocked, "Will you actually compare each and every ring? It will take you ages to find my ring."

Radhe replied, "Do not worry sir. I will not need more than a couple of minutes. I will not use the scales more than four times."

Within moments the ring was returned and the man was relieved. Radhe's uncle smiled with a lot of affection for his nephew.

How did Radhe find the real (slightly heavier) ring without using the scales more than four times ?
(Note: The scales were used without any weights and there was no means of knowing the exact weight of whatever was placed on the scales. The scales were used only to compare whether a ring on one side was heavier/lighter than the ring on the other side)
(3 points)

5. Lucky Student

St. Peter's High School in Patna was a wonderful school. It had more than two thousand students and it was very well known for its quality of education. Almost every year, the school produced 100% results and hardly any student had failed in the many years' history of the school. One of their most brilliant students was the 9th grade math genius, Bharat.

Bharat was just fourteen years old and was among the youngest in his class. Yet, he was extremely passionate about numbers and he was seen more often than not, bugging teachers and seniors for math problems and puzzles. He had represented St. Peter's in many competitions and was nicknamed 'The walking calculator'. He was a decent boy and although most students and teachers felt he was like a buzzing bee, every one had lots of respect and regard for his talents.

It was the 15th of August and it was the much awaited 72nd Independence day celebrations of India. The entire school was in a festive mode. There was a lot of extra care taken to keep the school premises clean as the chief guest, a retired scientist, was a well known person who had won tons of accolades and patents throughout his career. The principal was very particular about creating the right image when a man of such high importance was visiting the school.

It was soon time for the chief guest's speech. Everybody was excited and all the younger children had been warned to maintain silence throughout the lecture. More than two thousand pairs of hands clapped in unison as the elderly scientist walked to the podium to present his speech.

His speech was extremely interesting. He spoke about math, science, the universe and also of God. He told them, "Do you know, the universe is such a great creation? The precision is so exact that the probability of its occurrence is smaller than correctly calling out a hundred coin tosses consecutively. Can you imagine any human being so lucky to do such a thing?"

Bharat's mind began racing. He was almost on the edge of his seat. He could not help but raise his hand as high as he could. The scientist noticed. "Yes my boy, do you wish to say something?" Most of the teachers and students threw angry looks at Bharat as if to say, 'Is it really necessary to interrupt him?'

Bharat spoke with confidence and clarity, "Good morning sir, I do not know about a hundred. But I can assure you, that there is at least one student here, in this very crowd, who can correctly call out ten coin tosses."
 The scientist was amazed. "Who could be so lucky? Ten consecutive coin tosses? How can you be so sure, my boy?" Bharat quickly told everybody how it was possible. Everybody applauded in unison and the scientist exclaimed, "Well, my boy, you are indeed a genius!"

What was the strategy suggested by Bharat such that a person could successfully call out ten consecutive coin tosses correctly? How was he so sure that someone from his school would be able to do it?
(4 points)

6. Crossing the Fatal Bridge

Sikander (along with his brother, his dad and his grandfather) was in grave danger. It was dark and all of them were yet to reach home. They were businessmen and today there were too many customers to deal with. It was almost 9 pm when they reached the Fatal Bridge.
It was called Fatal Bridge because, well, it was fatal. The bridge was just a few minutes away from their home and it was very creaky. Although nobody had ever fallen from it, the very look of it was treacherous.

Sikander and his family were not just businessmen. They were cunning as well, very well versed in the art of conning. They often made money out of their clients and took advantage of almost every one.

Today had been their lucky day. They had stolen diamonds from their rivals with the help of a spy. They had the diamonds with them and that was another reason for them to be scared and more careful than usual. There had been many reports of thefts in the past week. And 9 pm had been declared as the curfew time for the people living in the vicinity. Exactly at 9 pm, a police patrol would cross under the bridge and would conduct a thorough search of whoever they saw crossing the bridge . It was already 8:41 pm when Sikander and his family reached the Fatal Bridge.

Since the four of them were regular users of the bridge, they knew their speeds. Sikander needed exactly one minute to cross the bridge; his brother needed two minutes; his dad was slower and required five minutes; the grandfather was the slowest of the lot and he crossed the bridge in no less than ten minutes. They had only one torch light and the bridge could bear the weight of only two of them at the same time.

Sikander quickly did the math and realized that they would just make it in time. Sikander crossed the bridge with his grandfather. The torch had to be used by both of them together, so Sikander could not walk at his normal pace and had to walk at his grandfather's speed. So it took them 10 minutes and they reached the other side at 8:51 pm (. Sikander came back with the torch in a minute (8:52 pm). He then crossed the bridge with his dad, taking them 5 minutes (8:57 pm). He came back with the torch in a minute (8:58 pm).
Thus, Sikander and his brother were relieved and began crossing the bridge, confident of reaching the other side by 9 pm.

However, the police arrived a minute earlier than expected. Both of them were caught in the middle of the bridge and soon there was a raid. The grandfather and father tried to escape, but could not see where they were going without the torch and were caught as well. The four of them were put on a boat. As the boat sailed along, Sikander and his brother suddenly broke lose and attacked the police. A fierce battle ensued and in the commotion, the bag of diamonds fell into the river. The thieves managed to steer the boat onto land and they ran helter skelter and within moments, the police lost track of them.

The police soon gave up the chase and Sikander and his family reached home, drenched and without most of their belongings (and also without the diamonds).
 As soon as they reached home, Sikander gasped at a sudden realization.

"What fools we were; if only we had crossed the bridge in a different way, we could have made it to the other side of the bridge even sooner. We could have reached the other side at 8:58 pm." The others took a moment or two to understand what Sikander was talking about. However, within moments all of them echoed his thoughts as well.

How could they have crossed the bridge in seventeen minutes (instead of nineteen) and reached at 8:58 pm (instead of 9 pm)?
(Note that only two can cross at a time and they need the torch to cross; so it has to be brought back after each crossing. Also, if two are crossing the bridge, the speed of the slower person will be considered since both of them need the torch.)
(3 points)

7. Pals on the plane

Jai Agnihotri boarded the flight. He was on his way to Paris and he was looking forward to having a pleasant, uneventful flight. He found his seat, stowed his luggage in the overhead compartment and was about to doze off when he noticed the man sitting to his immediate right. "Laxman!" he exclaimed. "Jai, what a pleasant surprise!" responded the gentleman instantly.

"My dear old mathematician, what a pleasure to see you after so long!" Jai replied, recollecting all the times in school when he had been overshadowed by his friend. "And you, my dear old artist, what are you painting next?" responded Laxman. Jai replied, "As of now I have to hope that God is painting a good future for me so that my work can sell and I can soon paint more and sell more! As they always say, the reward for hard work is more work." The two friends had always enjoyed each others' company since their childhood. Work commitments had been the main reason for their not meeting for quite a few years.

"How old are your kids now? You have three children don't you?" asked Jai matter-of-factedly and immediately regretted it.

Laxman replied, "Yes, I do have three children. The product of their ages is thirty six and the sum of their ages is today's date." Jai was stunned, "You expect me to solve that? Oh well, I should have known before asking a mathematician something."

Thus, Jai began working on the problem. However, he was not able to get the answer. He told Laxman, 'I am sorry, those clues do not seem to be enough. Can you give me another one?" Laxman instantly replied, "Oh, how foolish of me not to have mentioned. My youngest son loves to write and he seems to have a remarkable sense of logic. He also loves the colour blue."

Jai thought for a moment and soon produced the correct ages of the three children. "I must say, you are not as dumb as I thought," said Laxman and the two friends continued their banter all the way to Paris.

What are the ages of Laxman's children ?
(4 points)

8. Black and White on the Island

They were trapped! The evil king had imprisoned the ten of them on his luxurious island. Every year, the king punished those who spoke ill of him or did not obey his orders. The prisoners had to suffer for ten days on the king's luxury island. The king was also a huge lover of caps. He owned a huge collection of multi-coloured caps that he had accumulated over visits to various countries.

What made the king worse than a dictator was the way in which he loved to put the prisoners on the spot. Most of the times they were given the option to be set free in exchange for their friends and family; sometimes they were made to run blindfolded among snakes; and at other times the king personally whipped them.

But this year however, he had something else in store for them. He wished to play a game with his ten prisoners. A game that was worse than torture. He called it the 'Black and White cap' game, and it was his most recent invention (inspired by the many white and black caps that he received as a gift from his neighbouring kingdoms).
As the clock struck midnight on the tenth day of imprisonment, all the prisoners were called to meet the king. The king's minister read from a neat piece of parchment:

'Tomorrow at 9 am, all of you will play the 'Black and White cap game' in the presence of his highness. All of you will be made to stand in a straight line and hats of Black or White colors will be placed on your heads. You will only be able to see the caps of the persons in front of you. Your names will be called one by one, from back to front. As

your name is called, you have to guess the color of the hat on your head. You are not allowed to speak anything other than 'Black' or 'White'. Out of the ten of you, if more than one of you makes a mistake, your stay on this island will be doubled. However, if at least nine of you guess your hat colors correctly, you will all be set free tomorrow.'

The prisoners were excited and alarmed at the same time. This was something that they had not expected. If they somehow guessed the hat colors right, they would all be set free! Otherwise, the punishment was doubled. One of them exclaimed, "There is only half a chance that each of us gets it right and there are ten of us. It is almost impossible for more than five of us to get it right."
Suddenly, Mannu, a young farmer spoke up. He was a simple man but he had a very practical mind. He then suggested a method which gave lots of hope and confidence to the other prisoners.

They executed the strategy the next day, and much to the dismay of the king, the strategy worked and they were soon set free.

What was Mannu's strategy? Remember, only one person is allowed to be wrong. at least nine of them should correctly call out their hat colors.
(5 points)

9. Difficult Aliens

It was perhaps the most awaited day in the history of the Space Research Center. Life had been found on Mars and tomorrow, the 1st of January, 2047,would be the day where, for the first time in history, aliens would arrive on Earth. The entire Research Center was waiting for the moment with bated breath, along with the rest of the world.
 It was ten minutes before midnight when a message was sent to Earth from the Mars rocket:

Hello Earthians. This is the President of Mars speaking. We would like to test your intelligence before the beginning of an era of camaraderie. We have sent three alien robots, each representing a section of Martian society.

One robot always speaks the truth, representing the 'livito' section of Mars.
The second robot always speaks the untruth, representing the 'abrisa' section of Mars.
The third robot speaks truths and untruths at random and says whatever he likes, representing the 'camso' section of Mars.

All of them are programmed to answer to 'yes' or 'no' questions only.
All of them understand English of course but do not speak it.
They will chose to answer in our language using the words 'Si' and 'Pa', one of which stands for 'yes' and the other for 'no'.
You have to figure that out .
You can ask a total of three questions to the robots.

If you correctly guess their identities, Martians will step on Earthern soil for the first time in history.
However, if you fail to deduce the identities of our robots, we will return to Mars immediately and look for more superior beings to make friendships with.
Each of the three robots knows the identities of his companion robots.

Cheers,
Martian V, XCS.

The note ended and the Space Center was shocked with these sudden turn of events. 'What do we do?' was the common thought.

However, it was once more the brilliance of Mr. Pandey, that saved the day. He was the private detective of the Space Center and he had always been part of every case and mystery that had taken place in the last twenty years at the Space Center. He was extremely good in mathematics and most people felt he would have been better off as a mathematician than as a detective. He quickly told the others what questions had to be asked and a few minutes after midnight, the first aliens set foot on planet Earth.

What were the questions suggested by Mr. Pandey? (10 points)

(This problem is widely considered to be the toughest logical problem of all time and it took me over 90 minutes to solve it. Please do not be demotivated and I hope that you will do it in double quick time)

Total 37 points (a prime number and also my house number).

Answers

1. The question to be asked is: "Which is the road that leads to your home town?". Suppose the man at the sign board is from the city of Truth he will point towards his actual home town, the city of Truth, as he speaks only the truth. . However, if he is from the city of Lies, he will, nevertheless point towards the city of Truth , as he is a liar. So either way, he will have the same answer, the road that leads to the city of Truth, thus helping Prem.

2. Technically, there can be up to ninety-nine plants below the height of two feet. They can all be slightly less than two feet high as long as that one taller plant is a lot taller than two feet, thus compensating for all their shorter heights. Hence the average can be maintained even in the case of ninety nine plants having heights below two feet.

3. Until the realization happens, every one is under the impression that the other two are the only ones covered in soot and are laughing at each other. If Devilal is looking at both Veer and Avinash laughing, and if his face is clean, one out of Veer and Avinash would realize that Devilal and another person are laughing at him. Since, both Veer and Avinash continue laughing, it means that each of them is laughing at the other as well as at Devilal. Thus, Devilal realizes that even his face is covered in soot and he stops laughing.

4. The eighty one rings (80 + 1) in the jar are divided into groups of three (thus each group will be consisting of twenty seven rings). Two groups will be placed on the scales, one on each side. Suppose one side is heavier, it would mean that the heavier ring is in that particular group.

Suppose both groups have the same weight, it would mean that the heavier ring is in the third group. Thus, after the first weighing, Radhe reduces the size of the search to only twenty seven rings, i.e., the group of twenty seven which holds the heavier ring.

For the second weighing, these twenty seven are divided into three groups of nine rings each. Again, Radhe picks two groups and places them on either side of the scales. As before, he identifies the heavier group in just one weighing and reduces the size of the search to nine rings. For the third weighing, these nine are again divided into three groups of three rings each. Once again, Radhe picks two groups (of three rings each) and places them on either side of the scales. Again, in just one weighing, he finds the group that has the heavier ring. It is easy to locate the heavy ring from there. Ultimately for the fourth weighing, he has just three rings, that he divides into three groups of one ring each. He places a ring each on either side of the scales. If one side is heavier, that side holds the customer's ring. Instead, if both are equal in weight, the third ring is the customer's ring. Thus, merely four attempts are needed to discover the real ring from the eighty one as $3^4 = 81$

5. Let us assume we want to create a situation where a person correctly calls out four consecutive coin tosses. We would need sixteen people (2^4) for this experiment. Now we conduct a knockout tournament of coin tosses (each match will have two players who would each choose either 'head' or 'tail', and the winner goes to the next round) among those sixteen people such that in the first round (after one coin toss) eight people are eliminated.

Those eight people play the next round and four more get eliminated (after the second coin toss). By continuing this procedure, we arrive at the finals where two players participate. Each of them would have correctly called out three coin tosses to reach the finals. Hence, the one who correctly calls out the coin toss in the finals, wins the tournament and also would have successfully guessed four consecutive coin tosses correctly. Thus, for 4 coin tosses, we need 2^4 people for the knockout tournament. In general, the number of people has to be 2^n where 'n' is the number of coin tosses that have to be correctly called.

At St. Peter's , to create ten consecutive correct coin toss guesses, 2^{10}, i.e., 1024 people are needed. Since, the school has more than two thousand students, Bharat was very confident about this knockout tournament strategy working. So, the concept is all about creating a scenario where at least one person from the 1024 would be correctly calling out ten coin tosses (there *has* to be somebody who wins the tournament).

6.

First set : Sikander and brother (2 min) (8:43 pm)
Sikander comes back (1 min) (8:44 pm)
Second set : Dad and grandfather (10 min) (8:54 pm)
Brother comes back (2 min) (8:56 pm)
Third set : Sikander and Brother (2 min) (8:58 pm)
Total : 17 min.

7. Firstly, let us list all the combinations of three natural numbers that give us thirty six as the product; these are (1, 1, 36), (1, 2, 18), (1, 3, 12), (1, 4, 9), (1, 6, 6), (2, 2, 9), (2, 3, 6) and (3, 3, 4). The table below has tabulated these along with their sum in the bottom row:

1	1	1	1	1	2	2	3
1	2	3	4	6	2	3	3

36	18	12	9	6	9	6	4
38	*21*	*16*	*14*	*13*	*13*	*11*	*10*

Jai knows the date. We do not. The date can not be 38, so the first possibility is ruled out. However, if the date was any out of 21, 16, 14, 11 or 10, he would instantly have given the answer. He is confused because there is ambiguity; which means there are two combinations with the same total. Hence the date has to be 13! This leaves us with two possibilities: 1, 6, 6 and 2, 2, 9. Laxman tells him 'My youngest son...'. Thus, in the final answer, there has to be a *youngest*. '2, 2, 9' does not have a youngest. Hence, it has to be '1, 6, 6'.

8.

The person standing at the very back has to count how many hats of each color he sees in front of him. Suppose there are an even number of black hats, he says 'Black''. Suppose there are an odd number of black hats, he says 'White'.
The next person counts the hats of ahead of him and concludes, based on the first person's call, what color his/her hat is. Thus, the process continues, with each person keeping track of the colors of the hats of the people behind them, and remembering the clue given by the first person. Hence, each person looks ahead of them and based on whether there are even or odd number of black or white hats, decides what the color of his/her hat should be.
The last person (who is the one standing at the very front) does not see anything ahead of him. He would just have to count the number of hats mentioned by every person from the second person to the second last person, and decide,

based on the call of the first person, the color of the hat which lies on his head.

This whole process can be carried out by interchanging black with white or even with odd. Hence, only one person could be wrong in their call, i.e, the first person. Everybody else will get their calls correct. The king has said at least nine correct calls; hence this strategy works.

9. I will not reveal the answer to this legendary question. There are many answers possible. If you have gotten an answer and would like to grade yourself, I would suggest you to try it yourself and if it convinces you, why not? After all, there is no better judge of thy than thyself! If you are definitely sure about your answer or are very curious to knows, please drop me an e-mail and I will try my best to reply at the earliest.

EPILOGUE

Thus we end our brief journey. Hope you enjoyed the tour of this wonderful topic. This book is a humble effort of mine and I sincerely apologize for any unintentional mistakes. I hope this book is successful in planting the seeds of mathematical thoughts in your minds.

It is not as bad as you think it is. It is a misunderstood topic, and over the course of time many people have decided to hate it without any valid reason. I desperately felt that had to be changed.

I am not in anyways degrading the current system. I, as a student and 21st century person, have voiced what I felt, and shared what my experiences have taught me.

Math is not a subject that can be explained in one book. Thus, I have missed out on more than what I have presented.

I sincerely thank all my family members (mom, dad, sister, grandmother, uncles, aunts, cousin) and other well wishers who have constantly supported me. I would not even be in the position to write a book if not for the many guidances that I have received. Many wonderful teachers have taught me over the course of my life and I am indebted to them for making me think.

Thank you, reader, for reading this humble work of mine. Hope it benefits you! Ciao!

About the Author:

Born on 13th of March, 1999 to Sri Narasim Bhat and Smt. Shashikala Bhat, Raghavendra has been sharing his fun and passion for mathematics all over the world since the tender age of 13. He has performed well over 50 times in India as well as the US. Inside India he has performed at TedX Manipal, 2012; at NITK Engineer 2013, at IIT Chennai, at CMR Institute Bangalore, at International Algebra Conference at MIT, Manipal and also in various schools and colleges.

His shows at the National Conference on Numerical Methods and Modern Alegebra in 2013 been applauded by distinguished and senior mathematicians like Prof. E. Sampath Kumar, founder member of the Ramanujan Mathematical Society.

In April 2016 and September 2017, Raghavendra toured the US, giving his talks everywhere. He performed at prestigious places such as Harvey Mudd College, Wright State University, UIUC, and Miami University. He also performed at Intel.

His 90 minute show at PPC Udupi, was witnessed by Udupi's Admar Mutt swamiji, who blessed him a "true genius who has innate knowledge not taught by anyone". When in Grade 7, Raghavendra was recognized as a gifted child both in Math and Science by Duke University's TIP (Talent Identification Program). GT Foundation, Mangalore honoured him in December 2013. He has secured 2nd rank in the state in the science aptitude test, when in grade 7 and secured 6th rank in the country, when in grade 9. American Math Wiz and one of the top TED speakers, Dr. Arthur Benjamin, invited him to his house and Raghavendra spent a week with him in March 2016. Raghavendra has studied tabla under Pandit Madhav Achar in India. He has also taken advanced tabla classes from Pt. Sadanand Naimpalli, Mumbai. He has bagged many prizes for his tabla performances.

Before mathematics, Raghavendra was passionate about the universe and made large scale models of the planets and stars, and, when he was in 5th grade, presented to science teachers of his region.

The author at Princeton, September 2017, with legendary mathematician Prof. John Conway (2nd from right)

With world famous mathemagician, Dr Arthur Benjamin

With genius Mathematician from UCSD, Prof. Kiran Kedlaya